# Applicability of Reliability-Centered Maintenance in the Water Industry

Awwa
Research
Foundation

## About the Awwa Research Foundation

The Awwa Research Foundation (AwwaRF) is a member-supported, international, nonprofit organization that sponsors research to enable water utilities, public health agencies, and other professionals to provide safe and affordable drinking water to consumers.

The Foundation's mission is to advance the science of water to improve the quality of life. To achieve this mission, the Foundation sponsors studies on all aspects of drinking water, including supply and resources, treatment, monitoring and analysis, distribution, management, and health effects. Funding for research is provided primarily by subscription payments from approximately 1,000 utilities, consulting firms, and manufacturers in North America and abroad. Additional funding comes from collaborative partnerships with other national and international organizations, allowing for resources to be leveraged, expertise to be shared, and broad-based knowledge to be developed and disseminated. Government funding serves as a third source of research dollars.

From its headquarters in Denver, Colorado, the Foundation's staff directs and supports the efforts of more than 800 volunteers who serve on the board of trustees and various committees. These volunteers represent many facets of the water industry, and contribute their expertise to select and monitor research studies that benefit the entire drinking water community.

The results of research are disseminated through a number of channels, including reports, the Web site, conferences, and periodicals.

For subscribers, the Foundation serves as a cooperative program in which water suppliers unite to pool their resources. By applying Foundation research findings, these water suppliers can save substantial costs and stay on the leading edge of drinking water science and technology. Since its inception, AwwaRF has supplied the water community with more than $300 million in applied research.

More information about the Foundation and how to become a subscriber is available on the Web at **www.awwarf.org**.

# Applicability of Reliability-Centered Maintenance in the Water Industry

Prepared by:
**Christopher Fynn**, **Marius Basson**, **Steve Sinkoff**,
**Alastair Moubray**, and **Rick Nadeau**
New Dimension Solutions
734 Franklin Avenue #514
Garden City, NY 11530

Sponsored by:
**Awwa Research Foundation**
6666 West Quincy Avenue, Denver, CO 80235-3098

Published by:

Awwa Research Foundation

American Water Works Association

IWA Publishing

## DISCLAIMER

# CONTENTS

# TABLES

# FIGURES

# FOREWORD

The Awwa Research Foundation (AwwaRF) is a nonprofit corporation that is dedicated to the implementation of a research effort to help utilities respond to regulatory requirements and traditional high-priority concerns of the industry. The research agenda is developed through a process of consultation with subscribers and drinking water professionals. Under the umbrella of a Strategic Research Plan, the Research Advisory Council prioritizes the suggested projects based upon current and future needs, applicability, and past work; the recommendations are forwarded to the Board of Trustees for final selection. The foundation also sponsors research projects through an unsolicited proposal process; the Collaborative Research, Research Applications, and Tailored Collaboration programs; and various joint research efforts with organizations such as the U.S. Environmental Protection Agency, the U.S. Bureau of Reclamation, and the Association of California Water Agencies.

This publication is a result of one of these sponsored studies, and it is hoped that its findings will be applied in communities throughout the world. The following report serves not only as a means of communicating the results of the water industry's centralized research program but also as a tool to enlist the further support of the nonmember utilities and individuals.

Projects are managed closely from their inception to the final report by the foundation's staff and large cadre of volunteers who willingly contribute their time and expertise. The foundation serves a planning and management function and awards contracts to other institutions such as water utilities, universities, and engineering firms. The funding for this research effort comes primarily from the Subscription Program, through which water utilities subscribe to the research program and make an annual payment proportionate to the volume of water they deliver and consultants and manufacturers subscribe based on their annual billings. The program offers a cost-effective and fair method for funding research in the public interest.

A broad spectrum of water supply issues is addressed by the foundation's research agenda: resources, treatment and operations, distribution and storage, water quality and analysis, toxicology, economics, and management. The ultimate purpose of the coordinated effort is to assist water suppliers to provide the highest possible quality of water economically and reliably. The true benefits are realized when the results are implemented at the utility level. The foundation's trustees are pleased to offer this publication as a contribution toward that end.

Walter J. Bishop
Chair, Board of Trustees
Awwa Research Foundation

Robert C. Renner, P.E.
Executive Director
Awwa Research Foundation

# ACKNOWLEDGMENTS

The authors of this report are indebted to the following water utilities and individuals for their cooperation and participation in this project:

KEY INDIVIDUAL CONTRIBUTORS

- John Colbert, Asset Manager, Massachusetts Water Resources Authority
- Robert Lamb, Assistant Maintenance Superintendent, Denver Water Board
- Eugene Nicholas, Assistant Director, Broward County, FL
- Jean-Michel Seillier, Facilities Manager, Veolia Water Indianapolis

KEY CONTRIBUTING UTILITIES

- Broward County, FL
- Denver Water Board
- Massachusetts Water Resources Authority
- Veolia Water Indianapolis

We would like to acknowledge the advice and help of the Project Advisory Committee (PAC) including:

PROJECT ADVISORY COMMITTEE

- Jason Allen, Project Officer, AwwaRF
- Jennifer Warner, Project Officer, AwwaRF
- India Williams, Project Officer, AwwaRF
- John Colbert, Asset Manager, Massachusetts Water Resources Authority
- Allen Goodson, Manager, Operations and Maintenance, Planning, East Bay Municipal Utilities Department
- Alan Hess, Principal, Hess Consulting
- Parker Mitchell, Plant Superintendent, Charleston Public Works Department

The authors wish to acknowledge the assistance of Rick Nadeau and Paul Musca of Decima Research from Ottawa, Canada for their assistance in assimilating and presenting the data from the utility survey into meaningful figures and results.

# EXECUTIVE SUMMARY

## RESEARCH OBJECTIVES

The general objective of the research project has been to assess how water utilities can apply RCM* to new and existing infrastructure and to evaluate the costs and benefits of such a program.

In accordance with the AwwaRF RFP 2953, the research project addresses the following tasks:

1. Review of available Literature on RCM in the water industry
2. Identification of the components of an RCM program, using the SAE standard as a starting point, and how they could be applied in the water industry.
3. Identification of how utility maintenance programs are currently developed and if any utilities are currently using an RCM approach
4. Identification of how a utility would implement an RCM program, system-wide or on specific pieces of critical equipment (e.g. high-lift pumps, chemical feed systems) and the data requirements for same
5. Establishment of business rules that facilitate the identification of critical assets, regardless of the operating system being analyzed
6. Identification of the costs and benefits of implementing an RCM program (quantitative and qualitative)
7. Linkage of costs and benefits to drinking water utilities representing range of capacities and system configurations
8. Provision of a step-by-step procedure for implementing an RCM program identifying likely resource needs
9. Provision of Change Management recommendations to move from breakdown reactive approach to program built on reliability-centered principles

## APPROACH

To achieve the overall objective of the research project and address the specific project tasks required in the AwwaRF Request, NDS followed the approach detailed below (as outlined in the NDS research proposal).

- Completed two RCM pilot projects with two separate participating water utilities (Denver Water Board and Veolia Water Indianapolis) following the recommended approach described in Chapter 5 of this report, and carefully tracking costs and

---

* **Reliability-Centered Maintenance (RCM)** is a process used to determine the maintenance requirements of an asset in its current operating context – it is a scientific and systematic approach to ensure that the right maintenance task is done at the right time and that maintenance dollars are applied in the most effective manner. The process originated in the civil aviation industry where it has been instrumental in equipment reliability improvements of several orders of magnitude (100X) over the last 30 – 40 years. Over the last two decades RCM has been increasingly adopted over a wide-range of asset-intensive industries – including water and waste water utilities. (See **RCM Overview in Appendix D**).

benefits with detailed analysis of the "before" and "after" situations (to the extent of available information).

- Evaluated the RCM programs of two water utilities (Broward County, FL and Massachusetts Water Resources Authority) who have already completed a significant number of RCM analyses.
- Completed a substantial survey of water utilities to identify how water utilities are currently developing their maintenance programs and the current level of understanding and application of RCM practices.
- Conducted a literature review to assess the key literature available on RCM in general and specific to the water industry
- Compiled a comprehensive research report utilizing inputs from Items 1, 2, 3 and 4 above and addressing the AwwaRF Project Objectives and Tasks.

(The RCM process referred to throughout this proposal complies fully with SAE standard JA1011: "Evaluation Criteria for Reliability-Centered Maintenance Processes.").

## CONCLUSIONS

A number of conclusions have been reached as a result of this project:

### 1. Cost-benefit

The major conclusion of this report is that water utilities will typically derive extensive benefit from utilizing RCM to develop optimized maintenance programs for their asset base. This is well demonstrated in the evaluations of established RCM programs at the two water utilities and in the results achieved with the pilot projects performed as part of this research project. The caveat resulting from the survey and the experience of the authors of this report is that if projects are not properly structured or are approached informally at a low level in the organization, results may vary. This is indicated by the survey where those respondents who use RCM typically showed benefit but to varying extent. Detailed analysis of the established RCM programs in the case studies included in this report (at MWRA and Broward County) both showed substantial cost reductions while simultaneously improving asset performance. At the MWRA substantial cost savings in the order of **25-30%** were achieved by reduction of both Preventive and Corrective Maintenance hours while equipment reliability and plant availability was improved. The case study at Broward county also showed very significant maintenance cost reductions (>25%) from both reduced preventive and corrective maintenance while asset performance was improved.

The conclusion to be derived from the detailed evaluations on case studies of established RCM programs is that benefits from RCM substantially outweigh the costs of implementing the RCM program. The programs provided substantial payback from cost reduction aspects alone – before quantifying benefits of improved reliability and asset performance. *It should be stressed that the objective of an RCM program is to optimize the reliability and performance of the assets concerned by developing the most effective maintenance program. Cost-savings may result from an RCM program dependent on the prior situation but the primary objective of the program should NOT be to cut people or maintenance cost.*

## 2. Extent of usage of RCM

The survey indicates that RCM is currently utilized in a relatively small fraction of water utilities – less than 10% of the utilities responding to the survey.

## 3. Current maintenance program formulation tools

Most utilities that have developed maintenance programs have done so primarily from OEM recommendations and equipment history. The "Fixed-interval" maintenance tasks (also known as preventive maintenance tasks) are developed from several methods including OEM recommendations, equipment history, formalized preventive maintenance optimization processes, and RCM. When asked to specify the percentage that each of these methods have contributed to the development of the fixed-interval tasks at their utility, respondents on average assigned 48% to OEM recommendations, 31% to equipment history, 10% to formalized preventive maintenance optimization process, and 3% to RCM.

## 4. Strong and growing emphasis on managing physical assets at water utilities

Survey data revealed that there is an emergent focus by utility management on asset performance and optimal management of the utility's assets. Over half of respondents (57%) indicated that this focus is becoming an important issue for utility management, and 13% of respondents stated that optimal management and performance of the utility's assets is becoming ***a very critical issue for utility management***. Respondents were also asked to provide reasons for any change in management prioritization of physical asset maintenance and performance. When the first two reasons for this change in management prioritization were grouped together, aging equipment, facilities, or infrastructure (21%), the desire for better performance, higher efficiency, or optimization (16%), the need for cost reduction or to control costs (10%), and for compliance with regulatory requirements or more stringent regulations (10%) were the most commonly mentioned responses.

# RECOMMENDATIONS

The following recommendations are made:

## 1. Use an RCM process and RCM software compliant with SAE Standard JA1011

It is strongly recommended that utilities utilize an RCM process and supporting RCM software that is fully compliant with the SAE standard JA1011. This is a concise standard which lays out the seven steps of RCM which must be included and the information required at each step.

## 2. Adopt staged approach for RCM – pilot project, then large-scale rollout

The recommendation is for utilities to proceed with a two stage approach. The first stage (Stage 1) is to perform an RCM Pilot project (also known as Proof-of-Concept), applying RCM to several critical asset systems. Having developed a successful pilot and knowledge and confidence in the RCM process, the utility should then proceed with Stage 2 – Large-scale application of RCM. It is recommended in most cases to perform the large-scale application of RCM in order of asset criticality, based on an Asset Criticality Analysis which may be performed as part of Stage 1 or Stage 2.

**3. Utilize experienced RCM consultants to initiate program and provide training**

RCM has been applied extensively over the last twenty years in most asset-intensive industries, including the water industry. The successful application of RCM is now a science not an art. It is recommended that water utilities use the services of consultants specialized in RCM in the water industry to provide guidance and training in the initial stages of the project. Consultants can be used to help with the Pilot Program and to bring the utility up the learning curve to self-sufficiency. Experience indicates that RCM programs which flounder are those that do not follow the recommended approach in terms of how to structure a project, and perform the necessary change management and training. Reading a book about RCM is not usually an adequate preparation for success.

**4. Structure the RCM project to target measurable and achievable business goals**

An effective RCM project requires adequate funding and resources and must usually clear the normal ROI and payback hurdles to obtain the requisite funding as a business improvement project. To ensure funding and continued management investment, RCM projects must be directed at measured or specific business objectives and results must be carefully measured against these stated objectives.

**5. Use a review group approach**

It is a strong recommendation that RCM is pursued using a review group approach – for both existing and new assets. For existing assets this means that RCM analyses are conducted by a review group drawn from maintenance, operations and specialist personnel who know and work with the asset system under review as opposed to a single engineer or analyst doing the exercise in isolation. This is critical both from access to "the biggest pool of knowledge" perspective and the change management perspective. Analyses drawn up by one person in isolation are invariably flawed and are not well accepted by the stakeholders who maintain and operate the asset. For new assets, the review group constituents may vary but should involve process and design engineers, OEM specialists and such maintenance and operations personnel as may be available depending on circumstances.

**6. Invest in the recommended change management and RCM training**

The report lays out in detail what change management and training should be provided. Over time, RCM training should cover all maintenance, engineering and operations personnel plus senior management (in varying levels of time commitment and detail). It is a critical part of maintenance excellence and essential to the sustainability and strategic success of the program. The review group concept is a critical part of change management.

## 7. Address hidden failures

Hidden failures are those failures that are not evident during normal operations. Most of these are failures of protective functions (such as alarms, trips, pressure relief valves etc) which are provided to avert the consequences of the initial failure of the protected function (e.g. a boiler becomes over-pressurized). In modern industrial plant as much as 40% of the failure modes are hidden failures. Hidden failures are addressed in RCM processes compliant with the SAE standard. ***The recommendation is to make sure these are properly addressed in the RCM project***. Effective management of hidden failures is a vital component of reducing risk for the organization.

## 8. Set up capability to perform predictive technology (condition monitoring) tasks

The RCM process derives the optimal maintenance program – much of which is typically condition-based. This means that many of the fixed-interval tasks derived during the RCM process are **predictive** (condition-monitoring) tasks. These tasks monitor asset condition until an emerging failure is detected when an appropriate repair or replace task is initiated to restore normal asset condition.

The majority of predictive tasks involve the human senses but a significant and very important sub-set of these tasks are dependent on technology (e.g. vibration analysis, infra-red thermography, ultrasound, oil analysis etc). As a water utility moves down the path to world-class maintenance and a strong reliability program it is essential to invest in the predictive technologies and knowledge required to support the condition-based maintenance process and its associated condition monitoring requirements. This can be done on an incremental basis – adopting the most important technologies first and then building on this to greater sophistication. Dependent on utility size and organization it is desirable to create a small dedicated team focused on predictive technology.

# FUTURE RESEARCH

## Requirement for an Effective Maintenance / Work Management System

For utilities to derive the benefits of an optimized maintenance program developed using RCM they must have the capability to **systematically execute** the program of fixed-interval tasks according to the specified schedule and to act on the other recommendations of the RCM process (training, redesigns and other one-off actions). The capability to execute the maintenance program is predicated on an effective maintenance management structure/system. The overall maintenance management structure/system for the organization must be supported by a well-implemented Enterprise Asset Management system (EAM) or Computerized Maintenance Management System with well-defined business processes which are accepted and adhered to throughout the organization.

In our discussions with water utilities it is apparent that there is a wide range of capabilities and situations in this respect. Most utilities have some form of CMMS or EAM system but many do not. Among those utilities which do have CMMS or EAM systems there is a very wide range in terms of how they are used.

We strongly believe that this would provide a very productive area for research. The research topic could include the following:

- Current usage of CMMS and EAM systems among utilities
- Business functions covered
- Integration with back-office functions (materials management and financials/cost accounting/asset accounting)
- Support for Advanced Asset Management concepts and monitoring of asset conditions
- Percentage of maintenance work that is planned and scheduled in the EAM/CMMS system

# CHAPTER 1
# INTRODUCTION AND BACKGROUND

## OVERVIEW

This research project has been performed in response to AwwaRF RFP No. 2953.

The project is entitled "Applicability of Reliability-Centered Maintenance (**RCM**) to the Water Industry." **RCM (Reliability-Centered Maintenance)** is a methodology or process for developing the maintenance requirements for physical assets. It may be defined "as a process used to determine what must be done to ensure that any physical asset continues to do what its users want it to do in its present operating context" (Moubray 1997). Additional information and background on the origination of the RCM methodology is provided overleaf and in Appendix D.

## OBJECTIVES

The general objective of the research project has been to assess how utilities can apply RCM to new and existing infrastructure and to evaluate the costs and benefits of such a program.

In accordance with the AwwaRF RFP 2953, the research project addresses the following tasks:

1. Review of available Literature on RCM in the water industry
2. Identification of the components of an RCM program, using the SAE standard as a starting point, and how they could be applied in the water industry.
3. Identification of how utility maintenance programs are currently developed and if any utilities are currently using an RCM approach
4. Identification of how a utility would implement an RCM program, system-wide or on specific pieces of critical equipment (e.g. high-lift pumps, chemical feed systems) and the data requirements for same
5. Establishment of business rules that facilitate the identification of critical assets, regardless of the operating system being analyzed
6. Identification of the costs and benefits of implementing an RCM program (quantitative and qualitative)
7. Linkage of costs and benefits to drinking water utilities representing range of capacities and system configurations
8. Provision of a step-by-step procedure for implementing an RCM program identifying likely resource needs
9. Provision of Change Management recommendations to move from breakdown reactive approach to program built of reliability-centered principles

## SIGNIFICANCE OF THE PROJECT

The background to this research is the increasing emphasis on improved asset management in the water industry. Many water utilities are seeking to improve reliability, availability and utilization of their assets while also looking to control or reduce maintenance costs as they grapple with issues such as aging assets, more stringent environmental rules and tight budgets. In addition many utilities are poised to make significant investments in new infrastructure to replace aging assets, cater to increased user bases or adhere to stricter compliance requirements. This research project seeks to assess the benefits for water utilities in utilizing **RCM** to develop optimized maintenance programs which will improve asset reliability and avoid unnecessary cost – both for existing assets and new capital investments.

RCM is a process for deriving the optimal maintenance program for physical assets according to their current user requirements. RCM evolved from the aviation industry over thirty years ago in response to more technically complex systems, increasing maintenance costs and increasing downtime. The results of applying RCM in air transport have been a dramatic increase in safety, decreasing maintenance costs/flying hour and far higher equipment reliability and availability.

The air transport industry has learned to maintain its assets for high levels of reliability and service while doing so in a highly competitive, yet tightly regulated environment. The airline industry of the 1950's and 1960's experienced high costs from maintenance and many asset failures that led to disastrous consequences. The industry at that time had roughly 60 crashes per million takeoffs, of which more than two-thirds (forty+) were due to equipment failure. (That would equate to more than one large jet-liner crash occurring every day in terms of today's flight volumes!). Airlines tried what then seemed obvious – increase the maintenance so equipment failures could be avoided. Unfortunately the situation only deteriorated.

Due to the pioneering work of Nowlan and Heap in the 1960's and 1970's, the air transport industry came to the realization that traditional thinking would not work as larger, wide-bodied all-jet fleets of greater complexity containing far more components were introduced. Nowlan and Heap studied how assets failed and recognized that fundamentally different failures patterns were occurring, which required different approaches to avoid undesired consequences. They recognized that many failures were maintenance-induced, many were entirely random in nature, that many could be detected before they got to the totally failed state and that many were hidden under normal circumstances. Careful thought and analysis showed that all of these could be managed once they were understood and erroneous assumptions on failure put aside.

The airlines then developed a method or approach which fundamentally influenced the design of aircraft and also enabled the scientific determination of the type and amount of maintenance that decreased the failure consequences in terms of safety, the environment and operations. As a result of the widespread application of this method, they experienced a more than 30-fold reduction in crashes (today the rate is under two per million take-offs) and a 133-fold improvement in terms of crashes that are equipment–related (to less than 0.3 per million take-offs). The airlines now do far less of the fixed-interval type of overhaul work that was done previously. In those days 80% of aircraft maintenance was fixed interval work - today it is less than 20%. The work they do now is less expensive on a normalized basis. The method responsible for these step-change improvements is RCM.

RCM has now been industrialized and applied in virtually any field where physical assets are utilized. There have been a number of highly successful applications in the water industry with carefully documented results which has led to the commissioning of this research to communicate the facts about RCM to the broader water industry.

## RELIABILITY-CENTERED MAINTENANCE

The world of maintenance is in a state of flux as a result of greater automation and mechanization, increasing expectations, changing patterns of asset or equipment failure and new maintenance techniques and technologies. These factors have given rise to a growing need, felt not only by water utilities but also throughout industry, for a more effective and appropriate approach to maintenance.

A coherent review of maintenance strategies can only be undertaken by carrying out a zero-based review of the maintenance requirements of each item of equipment in its current operating context. This is because these requirements have changed dramatically in the light of modern thinking and a zero-based approach ensures that the maintenance program is developed on a scientific approach without any unfounded assumptions.

The evaluation of maintenance policies and the selection of maintenance tasks is a key aspect of maintenance management and most engineers do it continuously. However with the complexity of modern asset systems, the range of maintenance options and the number of additional factors to be considered in developing an optimized maintenance program it can no longer be done informally with any hope of producing a consistent, optimal and reliable outcome. RCM* solves this problem with a strategic framework that permits the maintenance program evaluation and selection process to be done consistently, quickly and confidently with an audit trail to explain exactly how and why the program was derived. To our knowledge, it is the only technique of its kind in existence, and it can lead to extraordinary improvements in maintenance performance wherever it is applied correctly. It does so in the following ways:

- RCM considers the consequences of each failure along with its technical characteristics. In doing so:
  - It integrates a review of operational failure consequences with the evaluation of safety and environmental hazards. This brings safety and the environment into the mainstream of maintenance decision-making
  - It continually focuses attention on the maintenance activities that have most effect on the performance of the plant. This means that every dollar spent on maintenance is spent where it will do the most good.
- RCM recognizes that different types of maintenance are applicable in different circumstances, and provides sound scientific rules for deciding which is most suitable in each situation. By doing so, it helps to ensure that the most effective forms of maintenance are chosen for each asset, and avoids the constraints and distortions that always follow the adoption of a single plant-wide maintenance policy.
- If RCM is correctly applied to existing maintenance systems, it typically reduces the amount of routine maintenance work (usually by 40% to 70%). On the other hand, if RCM is used to develop a new maintenance program, the fixed-interval workload that results is much lower than it would typically be if the system were developed by traditional methods.
- RCM was developed to help the air transport industry draw up maintenance programs for new types of aircraft before they enter service. As a result, it is an ideal way to develop such programs for new equipment, especially complex equipment for which little historical information is available. This saves much of the trial and error that is so

---

* An overview of the RCM process is included as Appendix D. This Appendix is provided courtesy of Aladon LLC and was written by John Moubray, a leading authority on the RCM methodology and author of the book RCM2.

often part of the development of new maintenance programs - trial that is frustrating and time-consuming and error that can be very costly or dangerous.

- RCM provides a common, easily understood technical language for everyone who has anything to do with maintenance. This gives maintenance, engineering and operations people a better understanding of what maintenance can (and cannot) achieve and what must be done to achieve it. This in turn improves both effectiveness and morale.
- An RCM review of the maintenance requirements of each item of plant in its operating context provides a firm basis for establishing labor policies and associated training programs, and for deciding what spares should be held in stock.
- RCM has been used successfully in a number of industries around the world. However, it is still new to industry in general, so companies which apply it now could enjoy a competitive advantage, at least to the extent that maintenance affects competitive standing.
- Although it is new to industry, RCM has been used for thirty years in what is probably the most demanding maintenance arena of all - civil aviation. This means that it has been tested and refined in the field to a much greater extent than any similar techniques. Increasingly RCM has a substantial successful track record in numerous other industries - including utilities (water, electric and gas).
- RCM can be implemented by a utility's own staff (after suitable training and initial program structuring), which reduces the need for expensive external services. Drawing maintenance, engineering and operations people into the decision-making process also means that all parties are more inclined to work together, and the results are more likely to endure.

What RCM cannot do is lift the reliability of any asset systems or items of equipment above the levels established by their design and the manufacturing processes that produced them. No form of maintenance can achieve this. However, RCM is exceptionally effective in helping management to achieve inherent reliability levels very quickly (hence its name). It also reveals when problems are beyond the scope of maintenance and where redesign should be considered.

For these reasons, RCM could become a central feature of equipment maintenance program development within water utilities in North America.

## SUMMARY

The above paragraphs suggest that if RCM is applied to all the assets used by water utilities, it could be of great value in terms of equipment reliability, safety, environmental integrity and maintenance cost effectiveness.

(For those not familiar with the RCM process, an overview written by leading RCM expert (John Moubray of Aladon LLC) is included as Appendix D along with a schematic of the RCM process and RCM decision tree diagram as Appendix E.)

# CHAPTER 2
# METHODS AND MATERIALS

## GENERAL APPROACH

To achieve the overall objective of the research project and address the specific project tasks required in the AwwaRF Request, NDS proposed to perform the following activities to carry out this research project, (as outlined in the NDS research proposal).

1. Complete two RCM pilot projects with two separate participating water utilities (i.e. Denver Water Board and Veolia Water Indianapolis) following the recommended approach described in Chapter 5, and carefully tracking costs and benefits with detailed analysis of the "before" and "after" situations (to the extent of available information).
2. Evaluate the RCM programs of two water utilities (i.e. Broward County, FL and Massachusetts Water Resources Authority) who have already completed a significant number of RCM analyses (as described in further detail below)
3. Complete a substantial survey of water utilities to identify how water utilities are currently developing their maintenance programs and the current level of understanding and application of RCM practices (as described in further detail below).
4. Conduct a literature review to assess the key literature available on RCM in general and specific to the water industry
5. Compile a comprehensive research report utilizing inputs from Items 1, 2, 3 and 4 above and addressing the AwwaRF Project Objectives and Tasks 1-9 specified in Chapter 1.

(The RCM process referred throughout this proposal complies fully with SAE standard JA1011: "Evaluation Criteria for Reliability-Centered Maintenance Processes.").

## TASK 1 - RCM PILOT PROJECTS WITH TWO SEPARATE PARTICIPATING WATER UTILITIES

In Chapter 5, we lay out the recommendations for water utilities to initiate Reliability Programs using RCM. These include recommendations on approach and a step-by-step list of recommended activities on how to proceed. One of the recommendations in terms of approach is that utilities proceed in stages using a **pilot project** or **proof-of-concept** stage to begin the program. The rationale for this is that the RCM process itself takes time and effort. It also embodies a great deal of new thinking, much of which is surprisingly counter-intuitive. As a result, it is wise for organizations new to this process to approach it in stages. This staged approach is built around the concept of pilot projects designed to enable the organization to learn:

- Exactly what RCM achieves in the water utilities business environment, and how quickly it achieves it. This in turn will enable utilities to decide, before committing to the large-scale application of this process, where the return justifies the investment.
- How to apply RCM correctly and safely from a purely technical point of view.
- What sort of people (skills) need to be involved directly in its application.

Bearing this in mind Task 1 of the research project consisted of performing two pilot projects with participating water utilities. These pilot projects were performed according to the recommendations laid out in Chapter 5 and have been written up as detailed reports which are included as Appendices F and G. The idea behind Task 1 was to give a detailed account of typical pilot projects, the level of effort, the deliverables produced, results achieved and perhaps most importantly the verdict or opinions of the participating utilities.

Task 1 addressed Objectives 2 and 4

## TASK 2 - RCM PROGRAM EVALUATION OF TWO WATER UTILITIES IN THE US

RCM programs have been successfully implemented in a number of water utilities across the world. Two of these water utilities based in the USA have expressed their willingness to share their experiences with their RCM programs within this research project. They are Broward County Environmental Operations Division based in Pompano Beach, Florida and the Massachusetts Water Resources Authority (MWRA) based on Deer Island Boston, Massachusetts.

Broward County has completed a number of RCM analyses on two water utility plants and one wastewater treatment plant. The MWRA has also completed a number of analyses on their Deer Island wastewater treatment plant and their water supply plants.

Task 2 consisted of an in-depth analysis of these projects and the writing up of a report on each detailing the approach, scope, activities, training and results achieved.

### RCM Program Evaluation Criteria

In Task 2, NDS carry out a detailed evaluation of both participating utility RCM programs using the following criteria as a guideline:

- What equipment or systems did the program cover?
- How was the equipment selected for analysis?
- To what extent was equipment duplicated?
- How many RCM meetings were needed for the analysis of each system?
- How many people at each level were directly involved in review groups?
- How many people were trained to implement the program?
- How many facilitators were trained and deployed?
- How many review groups did each facilitator work with at once (in other words, how many review projects were in progress at once)?
- What impact did the program have on existing resources?
- On what basis were the RCM reviews to be audited, and by whom in each plant section?
- What level of outside consulting support was required to initiate and sustain the program?
- How much did the program cost?
- What were the financial, operating and other benefits of the program?
- What level of management support did the program require?
- How long did it take to get the RCM recommendations implemented?

The results of these evaluations will then be complied into a detailed report.

## TASK 3 - WATER UTILITY MAINTENANCE PROGRAM DEVELOPMENT SURVEY

Task 3 was designed to find out what is really happening in water utilities in the US in terms of Maintenance Program and to what extent RCM is currently being used. NDS carried out a Maintenance Program Development survey which included an on-line questionnaire of a representative cross section of Water Utilities in North America. The survey describes the type and scope of the maintenance strategies deployed at these water utilities.

The results of Task 3 have been written up in a comprehensive report which is included in Appendix B.

Task 3 was focused on Objective 3 of the AwwaRF RFP for the RCM research project.

### Maintenance Program Development Questionnaire

NDS shall use the following questionnaire as a guideline:

- Are there formal programs in place to develop equipment maintenance programs?
- What maintenance strategy formulation techniques are used to develop equipment maintenance programs for critical physical assets?
- What maintenance strategy formulation techniques are used to develop equipment maintenance programs for non-critical physical assets?
- Are maintenance tasks for all equipment selected on their technical ability to address the cause of equipment failure and mitigate the consequence of failure?
- Are maintenance tasks for all critical equipment evaluated in terms of their worth doing criteria based on the consequence of failure?
- Has the plant equipment criticality analysis been determined?
- Are equipment failures and performance shortfalls reviewed to identify cause and appropriate failure management policies?
- Are equipment maintenance programs reviewed on a regular basis?
- Are responsibilities identified and assigned for the development of predictive and preventive maintenance programs?
- Are all practices for the maintenance program development formally documented?
- Does the utility have the management commitment towards reliability improvement initiatives?
- Does the utility have a reliability function defined?
- Does the utility use external/ internal benchmarking and networking to keep abreast of new initiatives?
- Are performance metrics defined for the activities of equipment maintenance programs?
- Are there training plans in place for individuals who develop equipment maintenance programs?

## TASK 4 - LITERATURE SURVEY

NDS has conducted a thorough literature survey on RCM as a general topic and application of RCM in the water industry. This literature survey covered books, papers and articles. The results are included in Appendix A.

The literature survey is designed to give those readers of the research report, who develop a deeper interest in RCM, ready access to the major literature contributions on RCM.

Task 4 addressed Objective 1 of the AwwaRF RFP.

## TASK 5 - RESEARCH REPORT

Task 5 consisted of compiling this comprehensive research report utilizing inputs from Tasks 1, 2, 3 and 4 above and addressing the AwwaRF Project Objectives and Tasks as specified in the AwwaRF RFP.

# CHAPTER 3
# RESULTS AND DISCUSSIONS

## INTRODUCTION

This chapter discusses the tasks performed during the project and results achieved. It covers the two pilot projects performed and the two cases studies completed at utilities with established RCM programs. It also includes summary discussion on the utility survey – a detail report of which is included as Appendix B.

## RCM PILOT PROJECTS

The RCM pilot projects were carried out at Denver Water Board and at Veolia Water Indianapolis (Indianapolis Water). Each of these pilot projects is reported on in detail in the Appendices - Appendix F (Denver Water Board [DWB]) and Appendix G (Veolia Water Indianapolis [VWI]).

The sections of the report covering recommendations to utilities draw heavily on these Pilot Project reports and the case studies of established, successful RCM programs.

## SUMMARY OF PILOT PROJECT 1 - DENVER WATER BOARD PILOT PROJECT

The RCM program was initiated at Denver Water by carrying out an RCM pilot project on asset systems (equipment) at Marston Pump Station.

Marston pump station is one of 18 pump stations owned and operated by Denver Water. The main function of Marston Pump station is to transfer water from the Marston treatment plant (at elevation 5503') to two reservoirs (Belleview and Kendrick) and to supply water to various distribution zones in the network. The primary source of water for the Marston pump-station is the Marston treatment plant with secondary supply from two other sources – the Foothills treatment plant (at elevation 5860') and the Moffat treatment plant (at elevation 5624'). Selection of feed source is performed manually by operations depending on treatment plant availability and load conditions.

Marston Pump Station was selected as a pilot project for the RCM program because of its importance in supplying water to the grid when the Foothills Reservoir is out of service. Foothills reservoir has been out of service on two occasions over the past two years for maintenance and repair work. With the Foothills Reservoir supply in full operation; Marston Pump Station becomes a secondary pump station – however, when the Foothills supply is down the Marston supply to the grid is then critical. The maintenance requirements are different for the two different operating conditions (i.e. when Foothills is in full operation or not). The asset systems (equipment) were selected using a criticality approach. NDS and DWB management reviewed the assets in the operating landscape, noted the consequences of failure and relative likelihood of failure of these systems, and then determined asset criticality. The assets were ranked, and the most critical selected for analysis.

The following asset systems were analyzed at Marston Pump Station during the RCM Pilot project:

- Low-Side and High-Side Pumps and associated controls and valves
- Pump Discharge Control Valves and associated controls
- Control Valve Hydraulic Pressure supply system
- HVAC (Cooling System)
- Boiler and Pump Station Heating
- Pump Station Drain and Sump Pumps
- Potable Water
- Pump Station Air supply
- Gas Engines
- Overhead Crane

In selecting the assets for analysis, the decision was made not to analyze both the Low-Side and High-Side Pumps due to a high degree of similarity. The Control Valves are also similar and the analysis results obtained from the Low-Side Pumps and Control Valves could readily be templated and applied to the High-Side Pumps.

As mentioned above, Denver Water Board has several pump stations, each with very similar equipment. Therefore, the asset systems (equipment) at Marston Pump Station were selected to be analyzed, with the concept that the Marston analyses could be adapted for pump stations with similar asset layout, appropriately adjusted for any differences in operating context.

**Approach Followed for Pilot Project**

The same approach was followed for both pilot projects - consistent with the standard Pilot Site approach detailed in Section 2 above.

The approach was to perform a pilot project at DWB consisting of RCM analyses of two sets of asset systems. NDS would manage the pilot project and facilitate the first set of analyses. NDS would also train a DWB Facilitator and then mentor the DWB facilitator who would facilitate the second set of analyses. The project approach is shown in the project plan included in Appendix F.

The pilot project was initiated with a 3-hour presentation to DWB stakeholders on November 25th, 2003. This was then followed up with the first 3-day course in December 2003. After this an NDS RCM consultant (Marius Basson) worked with DWB to plan the pilot projects in detail (Table 3.1).

The first set of RCM Analyses (those facilitated by NDS) tackled the following asset systems at Marston Pump Station:

- Low-side water pumps and isolation valves
- Hydraulic accumulator, pressure supply and ancillary equipment
- Control Valves (Ball and Cone type valves)

The first set of Analyses started on January 8, 2004 and was completed by January 23, 2004 (excluding the audit step). During this period the review group met for 20 meetings. The meetings were scheduled daily, from 7:30AM to 12:30PM and from 1:30PM to 4:30PM.

**Table 3.1**
**RCM pilot project outline**

| Item | Estimated task duration in days |
|---|---|
| Management Presentation | 1 |
| 3-day RCM Course | 3 Days |
| Pilot Planning | 3 Days |
| Facilitate RCM Analysis | 15 Full-time days |
| 3-day RCM Course | 3 Days |
| 10-day Facilitator Course | 10 Days |
| Facilitator Support | 10 Full-time Days |
| Project Support | 3 days |
| Compile Reports | 5 Days |

After completion of the first set of RCM Analyses (Set 1), in accordance with the proposed program, NDS then held a 10-day RCM facilitator-training course. This was held for both DWB and VWI. The AwwaRF plan provided for training one facilitator from each utility. However, based on management interest driven from the introductory RCM course and the initial analysis, DWB placed four Denver Water Board employees on the 10 day facilitator's course.

After completion of the facilitator's course DWB facilitators then facilitated the second set of analyses under the mentorship of NDS consultants. These started on April 5, 2004 and were completed by April 16, 2004.

The second set of analyses addressed the following asset systems:

- Marston Pump station chillers (coolers)
- Boiler and station heating
- Sump and sump pumps

After completion of the pilot project a presentation to DWB management was set up and delivered on May 21, 2004. At this presentation the Analyses were presented by the DWB facilitators to a group of managers of the appropriate facilities. This group consisted of the Operations and Maintenance Superintendents, Assistant Superintendents and Foremen of the plant. All members of the review group were also invited, and about ninety percent attended. The managers were provided with copies of the draft before the meeting.

### Pilot Project Results at DWB Marston Pump Station

The RCM Pilot project at Marston Pump station attracted significant management interest and generated some strong emotions. Perhaps the most significant outcome of the pilot project is that DWB management has since committed to the RCM process and devoted significant resources to additional RCM work. This management commitment/decision was based on the results achieved (Table 3.2) and the feedback from DWB personnel involved with the pilot project on their experience of working with RCM.

**Table 3.2**
**Summary of RCM results**

| Item | Number |
|---|---|
| Total Number of Functions | 109 |
| Total Number of Functional Failures | 125 |
| **Failure Mode Analysis** | |
| Total Number of Failure Modes Analyzed | 290 |
| Total Number of Induced Failure Modes (Failure Pattern F) | 57 |
| Total Number of Age-Related Failures (Failure Patterns A, B & C) | 32 |
| Total Number of Non-Age Related Failures (Failure Patterns D & E) | 152 |
| Total Number of Hidden Failures (Mostly Failure Pattern E) | 49 |
| **Failure Management Policies** | |
| Total Number of Predictive Tasks Specified | 25 |
| Total Number of Preventive Tasks Specified | 3 |
| Total Number of Failure Finding Tasks Specified (Hidden Failures) | 32 |
| Total Number of Compulsory Redesigns | 9 |
| Total Number of Recommended Redesigns | 54 |

**Table 3.3**
**Comparison of before and after situation by task type**

| Response to failure mode | After RCM | Before RCM |
|---|---|---|
| Predictive Tasks | 25 | 46 |
| Preventive Tasks | 3 | 13 |
| Failure-Finding Tasks | 32 | 5 |
| **Total fixed-interval tasks** | 60 | 64 |
| Run-to-Failure - No Fixed-Interval Maintenance | c. 230 failure modes | Not available |
| Redesigns | 63 | Not available |
| Failure Modes not managed | | 94 |

The period of review for the pilot project was of insufficient duration to do a detailed maintenance cost comparison for both corrective and preventive (fixed-interval) maintenance. (Cost comparison figures have been well analyzed and detailed for the two case studies which reviewed established programs which had been running for some time.)

The comparison shown in Table 3.3 analyzes the "Before" and "After" situation for the Fixed-Interval or Preventive tasks.

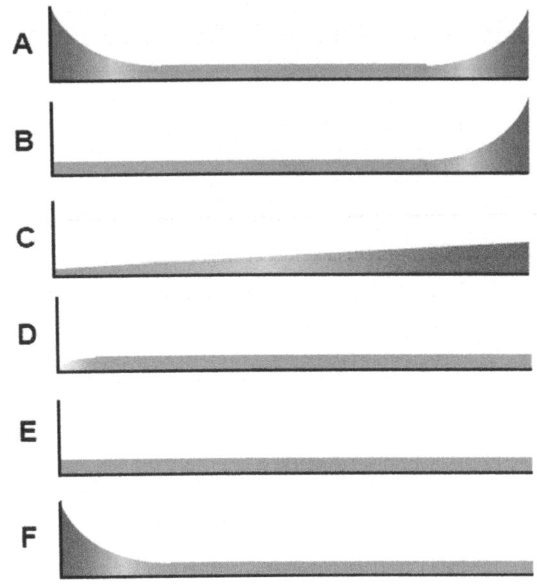


**Figure 3.1 Failure patterns and letter coding**

The failure patterns mentioned are shown in Figure 3.1. The categorization of the failure patterns and the letter coding was included in the original RCM work performed by Nowlan and Heap.

***Fixed-Interval (Preventive) Maintenance Program Analysis and Key Summary Points***

The number of fixed-interval (PM) tasks reduced overall by 6%. The composition of these tasks is however completely different and comprehensively covers all failure modes (including hidden). Although the reduction does not appear significant (when considered in aggregate) the analysis table (Table 3.4) and the following points illustrate specific details in the "before" and "after" situations which highlight the impact of the changes.

- The proposed redesign and modification of the boiler system (by replacing the boiler system with electrical heaters) reduced the number of semi-annual PM tasks by 21 tasks. This is a thirty-two percent (32%) saving of the current PM program. This will result in a saving of approximate 24 maintenance man hours/year resulting in a cost saving of $1080.00/year. This saving is continuous and can be multiplied by the number of similar installations if the proposed modification is accepted at all the Denver Water Board pump stations. A further saving manifests itself in not having to undergo the annual Boiler statutory inspection as required. Furthermore, the improvement in safety conditions stretches beyond any tangible savings. (It was found that one of the propane gas feed lines were color coded the same as the potable water line greatly increasing the risk that somebody could potentially cut into a live gas line)

**Table 3.4**
**Similarity analysis of before and after situation for RCM pilot**

| Similarity indicator | Indicator description | Number | % of new tasks |
|---|---|---|---|
| 1. RCM =Current | Tasks are the same | 12 | 20% |
| 2. RCM = Modified version of current | Same general failure management approach | 2 | 3.5% |
| 3. RCM specified a task for which no current task exists | Failure mode is not managed under current program | 14 | 24% |
| 4. RCM specified run-to-failure (NSM) where no current task exists | Similarity may be coincidental in most cases | 230 (APPROX) | |
| 5. Hidden failures not managed under current program | Hidden failures not identified in current maintenance strategy | 29 | 49% |
| 6. Current task exists where no failure mode is identified in RCM analysis or where the current task is not worth doing (based on RCM decision logic) | I.e. current maintenance tasks reduced or replaced by standard operating procedures | 19 | 29.5% |
| 7. Current tasks exists – RCM specifies different task | Preventive maintenance tasks changed to predictive maintenance tasks | 17 | 26.5% |
| 8. Current maintenance tasks no longer required following redesigns | Changing the operation or physical configuration of the assets removes requirement for the current maintenance strategy | 22 | 34% |

- RCM reduced the number of Pro-active maintenance tasks (Predictive and Preventive Tasks) significantly. The Predictive Maintenance tasks were reduced by twenty-one (21) and the recorded Preventive Maintenance Tasks were reduced by eleven (11). This relates to a reduction of 45% of the current maintenance program. The RCM analyses found that the maintenance tasks were mostly not worth doing since the current Pro-active maintenance tasks did not satisfy the criteria of being technically feasible and/or worth doing. Personnel are further performing unrecorded maintenance activities (such as roaming visits) that were found to be not worth doing. The savings resulting from the reduction of recorded maintenance tasks are approximately $ 3,047.00/year. The savings resulting in reduction of the unrecorded maintenance activities may exceed this number two-fold.
- It is also clear from the analyses that aging of equipment plays a small part in the overall reliability of the system and plant. Age-Related failures contributed only eleven percent (11%) of the total number of identified potential failure modes. In modern plants Age-Related failures are typically between eight and fifteen percent of the total potential failures recorded. It has been proven that time-based, fixed-interval

overhaul / replacement (time-based maintenance) has little effect on the overall reliability of the system(s) (also true for fixed-interval, usage-based e.g. hours, kms cycles etc). There are many items for which there is no effective form of fixed-interval maintenance.

- More significant than the Age-Related failures were the "Failure Pattern F" failures, failures dealing with human error and induced failures. Fifty-seven (57) potential failures were identified that satisfied the characteristics of "Failure Pattern F," the reversed J-curve. These failures are normally managed by a one-time change to one of the following:
  - To the physical configuration of the asset through redesign of the asset, or
  - To the capabilities of the maintainer and operator through training, or
  - To the operating and maintenance procedures through development of SOP documents
- The number of redesigns totaled sixty-four (64), including nine (9) compulsory redesigns.
- The most significant change in the two approaches (before and after RCM) lies in the management of hidden failures. The current maintenance program did not have any failure management policy in place for twenty-nine (29) hidden failures. Implementing appropriate failure management strategies for these hidden failures increased the maintenance activity by approximately fort-five percent (45%) from the current recorded maintenance activity. These failures normally deal with protective devices which fail in such a way that nobody knows that they are in a failed state during normal operating conditions. RCM identified forty-nine (49) hidden failures and developed failure management policies or failure-finding tasks (functional checks) for thirty-two (32) of the hidden failures. This is very significant in terms of the system reliability and improved integrity.
- Fixed interval hours under the RCM developed program are approximately 81 hours versus 150.5 fixed interval hours under the previous program and 730 non-fixed interval hours spent on roaming inspections.

### *Corrective Maintenance Key Points*

Not enough time has elapsed since the pilot project was performed to fully compare the impact on corrective maintenance hours. Early indications are that corrective hours will be reduced by the RCM pilot project recommendations when fully implemented.

### *Combined Fixed Interval and Corrective Maintenance Impact*

The revised fixed-interval maintenance program will deliver a substantial maintenance cost and operational cost saving when all the recommendations are implemented. The expectations are that implementation of this program will also result in substantial improvements in asset system reliability and integrity. This will lead to savings in downtime and corrective maintenance costs. The High-Side Pumps and Control Valves are exactly the same configuration as the Low-Side System. The savings that would follow from the completed Low-Side analyses will be visible in the High-Side system as well.

As noted the results of these analyses can be templated to other similar pump stations, taking into account any revisions in operating context. Thus savings demonstrated here will rapidly be extended across all DWB pump stations.

In summary, the RCM process has significantly reduced the number of fixed interval preventive and predictive tasks and associated resource requirements. The RCM process has instituted a significant number of functional checks to manage hidden failures which were previously ignored. Implementation of the RCM-developed maintenance program is anticipated to increase reliability and integrity and significantly reduce corrective maintenance.

**Results Summary for Denver Water Marston Pump Station Pilot Project**

- Strategic RCM program approved based on pilot side results and feedback.
- Preventive maintenance hours reduced from 150.5 hours to 81 hours.
- Effect on corrective maintenance cost and failures over extensive time period not yet determined - but no adverse consequences to date.
- Nine compulsory redesigns
- 29 hidden failures now managed under RCM program where no policy/task existed before
- Formalized, optimized maintenance program now in place which is auditable, defensible and developed using the proven scientific approach of RCM with decision process on how program was developed easily available to appropriate personnel.

## SUMMARY OF PILOT PROJECT 2 - VWI PILOT PROJECT

### Overview of RCM Project at Veolia Water Indianapolis

The AwwaRF pilot project was part of an overall proof-of-concept for an RCM program at VWI. The analyses that formed the AwwaRF pilot project for VWI were performed on assets at **Fall Creek Water Treatment Plant** and **Geist Water Treatment Plant**. These treatment plants are installations that are operated and maintained by VWI and are described in more detail below.

Additional context to the RCM proof-of-concept and pilot projects performed at VWI is provided in the section below. VWI is an innovative public-private partnership created to operate and maintain the water supply assets for the City of Indianapolis. This is a long-term contract where VWI is incentivized to manage the physical assets in the Indianapolis water supply system so as to optimize performance and life-cycle cost over the twenty year period of the contract.

***Veolia Water Indianapolis – Public –Private Partnership***

In 2002, the City of Indianapolis acquired the assets of the waterworks and selected Veolia Water North America to operate and manage the entire Indianapolis Water (IW) system, including its subsidiaries. This marked the start of the largest public-private water partnership in North America. With the creation of VWI, the City ensured that the waterworks system would receive the global expertise of Veolia Water while maintaining local control and oversight. The 20-year contract allows both the City and VWI to make investments in the waterworks infrastructure, technology, people and the community. The fee structure, including performance-based incentives and continual improvement is designed to promote a win-win theme – in everyone's best interest.

### *Partnership Incentives*

The City included 37 incentive criteria as part of the management agreement to provide outstanding service to its customers. VWI is responsible for operating towards achievement of those established criteria, identified in the following areas:

- Customer Service
- Water Quality
- Capital Projects
- **Technical, Operations and Maintenance (O & M)**
- Caring for the Community

### *Requirement for Maintenance Program and Management Systems*

The nature of the twenty year partnership with the City of Indianapolis (which commenced in 2002) meant that both VWNA and VWI had a strong interest in optimizing asset performance and minimizing asset life-cycle cost over the twenty year period of the contract. To this end VWI implemented the Synergen Enterprise Asset Management (EAM) system in 2003. At that time there was no formalized maintenance program in place at VWI. Having acquired and implemented the EAM system, VWI needed to develop an effective maintenance program for the asset base to ensure optimized performance and life-cycle cost over the twenty year operating agreement. VWI and VWNA management perceived that RCM could provide this need. The RCM pilot project was envisioned as a proof-of-concept for VWI to satisfy the requirement for a robust, consistent approach to develop an optimized maintenance program for the VWI asset base.

As no formalized maintenance program for the IW assets existed prior to the implementation of Synergen in 2003, this meant that there was no maintenance program of sound quality to load into Synergen. Based on inputs from utilities such as the MWRA and the experience of VWNA personnel with RCM (particularly Mr. Bill Fahey), VWNA and VWI decided to consider RCM as a solution for developing an optimized maintenance program. VWNA and VWI required a consistent, scientific methodology for developing maintenance activities and VWNA further required a rigorous program that could be deployed at other Veolia-operated water utility companies.

### *Fall Creek Water Treatment Facility – Location of Asset Systems for Analysis Set 1*

The first set of RCM analyses was performed on asset systems at the Fall Creek Water Treatment facility. This is a 32 mgd facility supplied by the Geist reservoir.

*Fall Creek Water Treatment Plant – Systems Selected*

The main asset systems at Fall Creek are listed below - the items highlighted in red were included in the RCM Pilot Project in the first set of analyses:

- Backwash Header
- **Filter beds [Included in Pilot Analysis set 1]**
- **Backwash water system [Included in Pilot Analysis set 1]**
- Surface wash water system
- Filter Inlet
- Filtered Water Outlet
- Pump Suction Header
- Finished Water Header
- Drainage
- Chemical Storage and Feed
- Chemical Feed Flushing and Dilution
- Sampling and analysis
- Sump Pumping
- HVAC
- Human Occupancy
- Safety
- Lighting
- Electrical Power
- Controls

*Geist Water Treatment Plant – Location of Asset Systems for Analysis Set 2*

Geist Water Treatment Plant treats well water from the aquifer downstream of Geist Reservoir. Plans are in place to extend the capacity of Geist Water Treatment Plant to double current capacity. Indianapolis Water personnel requested the RCM analyses to be conducted on existing equipment to prevent carry-over of design constraints and problems experienced with the current facility to the new extensions to be built

*Geist Water Treatment Plant Main Asset Systems*

- Wells
- **Well Pumps [Included in Pilot Analysis Set 2]**
- Station Inlet
- **Aerator / Reaction [Included in Pilot Analysis Set 2]**
- **Aerator / Reaction Outlet [Included in Pilot Analysis Set 2]**
- Filter Inlet
- **Pressure Filters [Included in Pilot Analysis Set 2]**
- Filtered Water Outlet
- Ground Storage
- Pump Suction Header
- High Service Pumping
- Wash water Pump

- Finished Water Header
- Drainage and Lagoon
- Sodium Hypochlorite Storage and Feed
- Ammonia Storage and Feed
- Fluoride Storage and Feed
- Bisulfite Storage and Feed
- Chemical Feed Flushing and Dilution
- Sampling and analysis
- Sump Pumping
- Air Compressor
- HVAC
- Human Occupancy
- Safety
- Electrical Power
- Controls

In selecting the assets for analysis, the decision was made to analyze the assets at Geist Water Treatment Plant that will be "duplicated" during the planned plant upgrade (i.e. **Well Pumps, Aeration Tank and Pressure Filters**). Two new wells are in the process of being commissioned and the Treatment Plant (pressure filters and backwash system) is to be expanded to include another 4 million gallons per day.

## Pilot Project Results at VWI Fall Creek Water Treatment Plant and Geist Water Treatment Plant

Tables 3.5 through 3.10 give a summarization of the results of the RCM analyses performed at VWI during the RCM pilot project.

### *Summary of RCM Analyses Included in Pilot Project*

***Geist - Well Pumps.*** Table 3.5 summarizes what was covered in the RCM analysis of the Geist Well pumps. No prior comparison is possible due to the absense of a formalized maintenance program prior to the RCM pilot project.

***Geist - Aeration Contact Tanks.*** Table 3.6 summarizes what was covered in the RCM analysis of the Geist Aeration Contact Tanks. No prior comparison is possible due to the absence of a prior, formalized maintenance.

***Geist - Pressure Filters.*** Table 3.7 summarizes what was covered in the RCM analysis of the Geist Pressure Filters. No prior comparison is possible due to the absense of a formalized maintenance program.

***Fall Creek - Sand and Gravel Filter.*** Table 3.8 summarizes what was covered in the RCM analysis of the Fall Creek Gravel Filter. No prior comparison is possible due to the absense of a formalized maintenance program prior to the RCM pilot project.

***Summary of All RCM Analyses for VWI Pilot Project.*** Table 3.9 aggregates the results of the RCM analyses perfomed as part of the Pilot project VWI. A significant amount of redesign items were discovered during theses analyses.

Figure 3.2 explains the Failure Pattern references used in Table 3.9.

**Table 3.5**
**Summary of RCM results - Geist well pumps**

| Item | Number |
|---|---|
| Number of Functions | 19 |
| Number of Functional Failures | 24 |
| **Failure Mode Analysis** | |
| Number of Failure Modes Analyzed | 73 |
| Number of Induced Failure Modes (Failure Pattern F) | 6 |
| Number of Age Related Failures (Failure Patterns A, B & C) | 19 |
| Number of Non-Age Related Failures (Failure Patterns D & E) | 28 |
| Number of Hidden Failures (Mostly Failure Pattern E) | 20 |
| **Failure Management Policies** | |
| Number of Predictive Tasks Specified | 17 |
| Number of Preventive Tasks Specified | 1 |
| Number of Failure Finding Tasks Specified (Hidden Failures) | 5 |
| Number of Compulsory Redesigns | 4 |
| Number of Recommended Redesigns | 23 |

**Table 3.6**
**Summary of RCM results - Geist aeration contact tanks**

| Item | Number |
|---|---|
| Number of Functions | 20 |
| Number of Functional Failures | 26 |
| **Failure Mode Analysis** | |
| Number of Failure Modes Analyzed | 70 |
| Number of Induced Failure Modes (Failure Pattern F) | 10 |
| Number of Age Related Failures (Failure Patterns A, B & C) | 13 |
| Number of Non-Age Related Failures (Failure Patterns D & E) | 40 |
| Number of Hidden Failures (Mostly Failure Pattern E) | 7 |
| **Failure Management Policies** | |
| Number of Predictive Tasks Specified | 5 |
| Number of Preventive Tasks Specified | 11 |
| Number of Failure Finding Tasks Specified (Hidden Failures) | 4 |
| Number of Compulsory Redesigns | 0 |
| Number of Recommended Redesigns | 10 |

**Table 3.7**
**Summary of RCM results - Geist pressure filters**

| Item | Number |
|---|---|
| Number of Functions | 8 |
| Number of Functional Failures | 12 |
| **Failure Mode Analysis** | |
| Number of Failure Modes Analyzed | 38 |
| Number of Induced Failure Modes (Failure Pattern F) | 6 |
| Number of Age Related Failures (Failure Patterns A, B & C) | 11 |
| Number of Non-Age Related Failures (Failure Patterns D & E) | 20 |
| Number of Hidden Failures (Mostly Failure Pattern E) | 1 |
| **Failure Management Policies** | |
| Number of Predictive Tasks Specified | 4 |
| Number of Preventive Tasks Specified | 6 |
| Number of Failure Finding Tasks Specified (Hidden Failures) | 0 |
| Number of Compulsory Redesigns | 0 |
| Number of Recommended Redesigns | 4 |

**Table 3.8**
**Summary of RCM results - Fall Creek sand and gravel filter**

| Item | Number |
|---|---|
| Number of Functions | 33 |
| Number of Functional Failures | 53 |
| Failure Mode Analysis | |
| Number of Failure Modes Analyzed | 203 |
| Number of Induced Failure Modes (Failure Pattern F) | 23 |
| Number of Age Related Failures (Failure Patterns A, B & C) | 102 |
| Number of Non-Age Related Failures (Failure Patterns D & E) | 44 |
| Number of Hidden Failures (Mostly Failure Pattern E) | 34 |
| Failure Management Policies | |
| Number of Predictive Tasks Specified | 60 |
| Number of Preventive Tasks Specified | 26 |
| Number of Failure Finding Tasks Specified (Hidden Failures) | 10 |
| Number of Compulsory Redesigns | 10 |
| Number of Recommended Redesigns | 19 |

**Table 3.9**
**Summary of all RCM analyses performed in VWI pilot project**

| Item | Number |
|---|---|
| Total Number of Functions | 80 |
| Total Number of Functional Failures | 115 |
| **Failure Mode Analysis** | |
| Total Number of Failure Modes Analyzed | 384 |
| Total Number of Induced Failure Modes (Failure Pattern F) | 45 |
| Total Number of Age Related Failures (Failure Patterns A, B & C) | 145 |
| Total Number of Non-Age Related Failures (Failure Patterns D & E) | 132 |
| Total Number of Hidden Failures (Mostly Failure Pattern E) | 62 |
| **Failure Management Policies** | |
| Total Number of Predictive Tasks Specified | 86 |
| Total Number of Preventive Tasks Specified | 44 |
| Total Number of Failure Finding Tasks Specified (Hidden Failures) | 19 |
| Total Number of Compulsory Redesigns | 14 |
| Total Number of Recommended Redesigns | 56 |

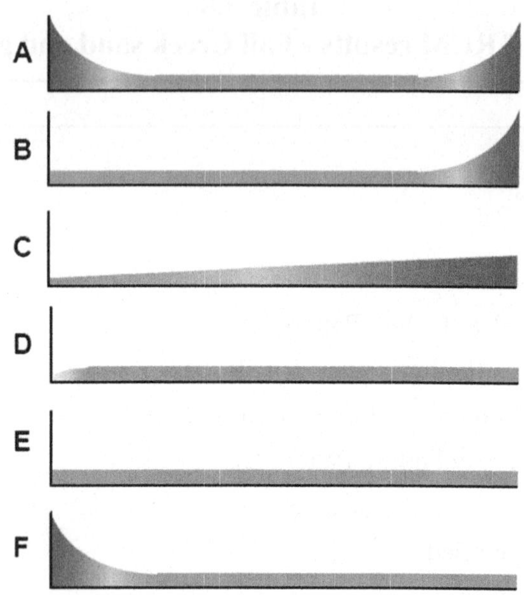


**Figure 3.2 Failure patterns and letter coding**

**Table 3.10**
**Before and after comparison by task type**

| Response to failure mode | After RCM | Before RCM |
|---|---|---|
| Predictive Task | 86 | Not available |
| Preventive Tasks | 44 | Not available |
| Failure Finding | 19 | None |
| Run-to-Failure - No Fixed-Interval Maintenance | 219 approx | Not available |
| Redesigns | 70 | - |
| Failure Modes not managed | | 16 (separate analyses) |

### *Comparison of Before and After Situation by Task Type*

Table 3.10 gives the normal before and after summarization. Tha lack of a prior fomalized maintenance program limits the opportunity for valid comparison.

### *Fixed-Interval Maintenance Key Summary Points*

As noted in the introduction, VWI implemented the Synergen EAM system in 2003 to manage the VWI assets and the execution of maintenance work. There was however no existing formalized maintenance program in place that could be loaded into Synergen from the legacy systems. This meant that from the perspective of the RCM pilot projects, there was essentially no reliable, comparable "before" information that could be obtained with respect to the maintenance program prior to RCM. Anecdotally, the RCM review group members were of a consensus in indicating that the RCM-developed program specified less fixed-interval maintenance than what was being done previously based on various informal systems in place. However this perception could not be verified due to the lack of an existing formal system of preventive maintenance.

- The key point in relation to the fixed-interval maintenance program is that after the RCM analysis for these asset systems, VWI is now in possession of a valid, optimized maintenance program for these assets - which is "owned" and agreed to by both Maintenance and Operations and is ready for loading into the Synergen system (once signed off by management) for systematic execution according to the specified frequency.
- It is noted that RCM typically reduces the number of *pro-active* maintenance tasks (Predictive and Preventive Tasks) significantly - channeling maintenance resources into the most effective forms of maintenance. This reduction of the fixed interval program is usually between 40-70%. In the review group discussions, the RCM process was seen to eliminate numerous existing maintenance tasks that did not satisfy the criteria for "technically feasible" and/or "worth doing." It was also established that personnel were regularly performing unrecorded "condition monitoring/inspection" maintenance activities (such as roaming visits) that were found to have little value and to be not worth doing.

- "Failure Pattern F" failures. These are failures dealing with human error and induced failures i.e. failures that show a high infant mortality. Forty-five (45) potential failures were identified that satisfied the characteristics of "Failure Pattern F i.e. the reversed J-curve. These failures are normally managed by a one-time change to one of the following:
  - To the physical configuration of the asset through redesign of the asset, or
  - To the capabilities of the maintainer and operator through training, or
  - To the operating and maintenance procedures through development of SOP documents
- The number of redesigns totaled seventy (70), including fourteen (14) compulsory redesigns.
- The most significant change in the two approaches (before and after RCM) lies in the management of hidden failures. The current maintenance program did not have any failure management policies in place for nineteen (19) hidden failures that were found to require failure-finding tasks. In contrast to the preventive and predictive fixed-interval tasks mentioned above, the RCM process here increased the fixed-interval maintenance activity from the current maintenance situation. These hidden failures normally deal with protective devices that fail in such a way that nobody knows that they are in a "failed state" during normal operating conditions. The RCM process identified sixty-two (62) hidden failures and developed failure management policies or failure finding tasks (functional checks) for nineteen (19) of the hidden failures. This is significant in terms of system reliability and improved integrity and managing the organization's risk profile.

### *Corrective Maintenance Key Points*

Insufficient information is available at this time to quantitatively compare the impact of the RCM analyses on the amount, cost of or downtime related to corrective maintenance. However perceptions coming back from the site are positive in this respect i.e. less or similar levels of corrective maintenance are occurring after implementing the RCM-developed fixed interval program.

### *Combined Fixed Interval and Corrective Maintenance Impact*

As noted the RCM-derived fixed - interval program cannot be compared accurately with what was done historically at IW as there was no prior formalized program and not enough time has elapsed to really analyze impact on overall maintenance levels (both fixed-interval (preventive) and corrective). In summary it is too early to provide hard data at this site on cost and asset performance metrics. Both management and review groups have a high belief in the quality of the RCM analyses and the resulting program and one-off recommendations. There is a strong perception among management and the review group's members that over time the RCM process will minimize life-cycle costs while improving asset performance and reliability. This has resulted in VWI moving ahead with a larger-scale application of RCM as noted below. RCM has become the tool of choice at VWI for developing maintenance programs on critical/significant asset systems – and on assets in the design phase.

### *Benefits From RCM Pilot Projects at VWI*

1. Optimized, scientifically derived maintenance program in place for the assets. As noted above the key benefit from the RCM analyses is that VWI now has a well documented, auditable, optimized maintenance program in place for these assets – versus the informal reactive approach previously followed. This enables a living program and further optimization (to optimize any situation one must introduce systems, measurement and order). It is not possible to compare the program to the status quo as that was not formalized into a system. RCM certainly typically provides a leaner program than OEM recommendations. It must be noted that OEM recommendations are not available for many of the assets now operated and maintained by VWI due to the age of the assets. VWI has established a robust baseline program on these assets using a scientifically sound methodology which creates an easy-to-follow audit trail on how the maintenance program was developed and what decisions were made in doing so.

   The outcome of a well managed maintenance function executing an optimized maintenance program typically includes improved reliability (reduced occurrence of failures with undesirable consequences), reduced life-cycle cost and a high confidence level in asset performance. Additional notes on benefits from the RCM pilot program are made below. As mentioned previously quantitative comparison is limited by the lack of history and the limited operating time post RCM on these asset systems. However user feedback on the process to management and the quality of the analysis output has led to management committing to RCM as the process for developing maintenance programs for significant systems.
2. Greater safety and environmental integrity
   - The RCM analysis of the Gravity Filters will contribute to improved safety and environmental integrity by addressing those failure modes that have safety and environmental consequences, (and were not previously addressed), with maintenance tasks and SOP's to reduce the risk of injuring personnel and breaching the NPDES permit.
   - Targeted employee training will assist in reducing the number of safety incidents overall.
   - VWI reduced the recordable incident rate 22% in 2003 and 18% in 2004 when compared with the year previous. The recordable incident rate of 7.07 for 2004 was an all-time low for VWI.
3. Higher plant availability and reliability

### *Gravity Filter – FCTP*

The scope for Gravity Filter improvement depends clearly on the performance of the Gravity Filters at the outset of the analysis. The Gravity Filters have been in operation for almost 90 years and have performed reasonably well. Nonetheless RCM is expected to achieve significant improvements regardless of the starting point. Plant performance is of course improved by reducing the number and severity of unanticipated failures, which have operational consequences. The RCM analysis of the Gravity Filters will help to achieve this in the following ways:

- The systematic review of the operational consequences of every compressor failure mode which has not been dealt with as a safety hazard, together with the stringent criteria used to assess task effectiveness, ensure that only the most effective tasks have been selected to deal with each failure mode.

- The emphasis placed on on-condition (predictive) tasks helps to ensure the potential failures are detected before they become functional failures. This helps reduce operational consequences in three ways:
  - Problems can rectified at a time when stopping the machine will have the least effect on operations
  - It is possible to ensure all the resources needed to repair the failure are available before it occurs, which shortens repair time
  - Rectification is only carried out when the assets really need it, which extends the intervals between corrective interventions. This in turn means the asset has to be taken out of service less often.
- By relating each failure mode to the relevant functional failure, the information worksheet has provided a tool for a quick failure diagnosis, which leads in turn to shorter repair times.

***Greater Maintenance Efficiency (Cost-Effectiveness).***

***Gravity Filters.*** The RCM analysis of the Gravity Filters will help to reduce, or at least control the rate of growth of maintenance costs in the following ways:

***Less Routine Maintenance.*** The application of RCM to the Gravity Filters reflects a reduction in comparison to the perceived OEM routine maintenance workload. This reduction is partly due to a reduction in some of the OEM routine tasks, but mainly due to an overall increase in the intervals between tasks.

***Water Quality.*** Water quality is a critical issue for VWI and VWI has accepted the challenge of resolving significant taste and odor problems related to algae growth in the drinking water supply reservoirs. Ensuring correct operation of the filtration and odor removal asset systems through a scientifically designed maintenance program is key to ensuring this objective is met consistently.

### *Intangible Benefits From RCM Pilot Project*

***Knowledge of the Assets.*** The greatest intangible gain from the RCM2 program in the opinion of VWI management and facilitators was knowledge of the assets on the part of operators and maintainers. Working through the details of each system in what is an old plant resulted in all group members learning much about the inner workings of the systems they have worked with for many years, as well as a better view of how each fits into the plant. The gains are illustrated by the positive attitude and support for the program by VWI employees.

***Communication and Teamwork.*** In the opinion of VWI management, the next most significant intangible effect of the RCM program was the improvement it generated in communication, both between Operations and Maintenance and between management and the workforce. RCM generated some anticipated initial suspicion from the maintenance group that the real intent behind the RCM2 initiative was staff reduction. This was eventually dispelled.

***Ownership of the Maintenance Program.*** One effect of the involvement of the "blue-collar" input in the development of maintenance strategies that the RCM2 process encourages, is a greater feeling of ownership on the part of the technicians who actually perform the maintenance.

***Drawings.*** Drawings were not available for some of the assets systems where RCM analyses were conducted. This led to frustration and uncertainties and illustrated the need for current drawings at the appropriate level of detail.

***Safety Benefits.*** Neither treatment plant had significant safety problems before the RCM2 program began. However, several analyses have produced significant recommendations to improve safety and environmental integrity. These include operating procedures that are not currently developed.

### *Outcome of RCM Pilot Program for VWI*

It has been noted that prior to 2003 there was no formalized maintenance system in place at Indianapolis Water and no formalized plant-wide maintenance program existed for VWI to utilize going forward for the twenty year operating and maintenance contract. Maintenance was essentially performed using informal systems at each plant based on the knowledge and experience of the personnel. Much of the maintenance was reactive in nature.

VWI undertook a proof-of-concept project to validate RCM as a process to develop an optimized maintenance program for the VWI asset base. The AwwaRF pilot project was part of this proof-of-concept.

As noted below, the outcome of the proof-of-concept is that RCM has been validated as a crucial component of VWI's approach to improving maintenance and asset performance during the life of the contract.

This excerpt is from the VWI annual report for 2004.

> VWI successfully implemented Synergen in September 2003. Synergen is an Enterprise Asset and Work Management system (EAM) / Computerized Maintenance Management System (CMMS) that allows companies to optimize asset performance and reliability. Through the use of the CMMS, maintenance programs are streamlined and the full life cycle of the asset is taken into consideration. Synergen is a web-structured program that allows projects to share information to improve overall efficiency.
>
> Maintenance costs were more accurately captured and reporting to the client was greatly improved. Preventive Maintenance (PM) tasks were generated based off calendar date and/or run time. Initially the PM tasks were developed from past practice, but with the implementation of Reliability Centered Maintenance (RCM), the frequency and scope of the PM are changing for improved asset management.
>
> RCM is a process used to determine what maintenance activities should be done to an asset to ensure it operates at the user's expectations throughout its life. RCM uses seven separate activities to identify what should be done to properly maintain an asset. VWI has analyzed over 20 separate processes throughout the overall system. Many predictive tasks have been established and the results have been favorable.
>
> *Excerpt from pg 7, Veolia Water Indianapolis Annual Report 2004*

In summary, the outcome of the RCM proof-of-concept work was that RCM was adopted as the Maintenance Program Formulation process by VWI for significant or critical assets. During the Pilot project VWI, acquired the ability to proceed with RCM in a largely self-sustaining capacity.

***Results Summary for Pilot Projects at VWI Fall Creek and Geist Water Treatment Plants***

- RCM adopted at VWI as maintenance program formulation policy for critical assets
- No quantitative comparison with previous situation due to lack of existing formalized preventive maintenance program
- Effect on corrective maintenance cost and failures over extensive time period not yet determined – positive feedback from site to date.
- Fourteen compulsory redesigns
- 19 hidden failures now managed under RCM program where no policy/task existed before
- Formalized, optimized maintenance program now in place which is auditable, defensible and developed using the proven scientific approach of RCM with decision process on how program was developed easily available to appropriate personnel.

## EVALUATIONS OF EXISTING RCM PROJECTS

Evaluations on existing RCM projects were carried out at Broward County in Florida and at Massachusetts Water Resources Authority in Boston, MA. These evaluations are covered in detailed reports which are attached as Appendix H (Broward County, FL) and Appendix I (MWRA).

### Summary of Broward County OES RCM Program

The RCM program began in 1998 when Eugene Nicholas, Assistant Director, attended a public course in RCM2. After returning from the course, Mr. Nicholas decided to motivate an RCM2 program at Broward County OES. A competitive RFP process was initiated for RCM services and from this process a project was launched to train the personnel necessary to perform RCM2 analyses on all significant equipment on the three plants under his authority.

The initial contract was for 12 pilot analyses, for which the consulting firm selected (Strategic Technologies Inc.) trained the necessary people, including the first three facilitators. When the pilots produced satisfactory results, the Broward County EOD decided to extend the RCM program to the rest of the facility. Two further facilitators were trained and another 3-Day RCM2 Intro course was run, to bring the number of trained people up to the 35 noted below. The external consulting firm also provided occasional RCM audits to ensure quality of the program.

The first assets to be analyzed were the Water Treatment [Softening] Unit from the District 2A Drinking Water Plant, the Park Road Interconnect from District 3A, and three Clarifier A units and Anaerobic Digester 1 from the Waste Water plant. These assets were selected on the advice of the RCM2 Consultants, as it was felt they would be a reasonably simple exercise for the newly trained facilitators.

After these, the program moved on to cover the following items:

### Waste Water Plant

- 4 RAS Modules (March/June 2000)
- 13 Clarifiers (June 2000)

- 16 Aeration Basins (October 2000)
- Chlorine Scrubber (April 2001)
- 6 Boilers (April/July 2001)
- 3 Electrical Buses (May/July 2001)
- Diesel and Electric Generators (July 2001)
- 3 Belt Presses (June 2001/January 2002)
- Ocean Outfall (January 2002)
- Conveyor Belt Press (February 2002)
- 6 Anaerobic Digesters (March 2002)
- Conveyor Belt Press Air Compressor (March 2002)
- Injection Well System (April 2002)
- 6 Dissolved Air Floatation Tanks (April/July 2002)
- Water Reuse System (February 2003)
- EOD SCADA (July 2003)

**District 1A Drinking Water Plant**

- Vac Filters (March/August 2000)
- 2 Slakers (August/November 2000)
- Chlorine Scrubber (January 2001)
- Ammoniator (January 2002)
- Thickener (July 2002)
- Ferric Chloride System (February 2003)
- 8 Filters (February 2003)
- Fluoride System (February 2003)
- Polymer System (February 2003)
- Water Treatment Unit (February 2003)
- SCADA (June 2003)

**District 2A Drinking Water Plant**

- Polymer System (March 2000)
- Ferric Chloride System (July 2000)
- Backwash System (March 2001)
- Chlorine Scrubber (April 2001)
- 6 Filters (March/October/November/December 2001)
- High Speed Pumping System (September 2002)
- Ammoniator (February 2003)
- Fluoride System (February 2003)
- 2 Slakers (February 2003)
- SCADA (April 2003)

**Well Fields / Others**

- 10 SRWF Wells (August 2001)
- 180 Retail Lift Stations (March 2002)
- 10 NRWF Wells (April/July/August 2002)
- 57th Street Interconnect (February 2003)

Management believes all significant equipment either has been or will be analyzed. The order the analyses were performed in was created by the Superintendents of each plant, on the basis of their experience with the assets. There was no formal criticality assessment process.

From the above lists, it is readily apparent that many analyses can be duplicated or used to create templates. At the time of writing, the RCM2 database contained 310 individual analyses for different equipment/asset systems as shown above. The 310 analyses cover 41 different technical system/asset types. This is obviously skewed by the Retail Lift Stations which account for 180 of the analyses, with the remaining 120 spread over 40 different asset types – as listed above.

In most cases, the analyses performed on each asset type have near-identical numbers of failure modes, suggesting that, as might be expected, there is little design difference between assets of the same type. This suggests that it was possible to make extensive use of templates when analyzing any of the plants involved, for assets of the same type.

After the audit team had signed off a new analysis, the group would reconvene with all appropriate drawings and work through the analysis to make the necessary changes for each asset of the same type. This usually took one meeting of three hours, in which time several analyses of assets of the same type could be made.

The database shows that, in most cases, one analysis was completed and audited, a process that usually took about two months, before assets of the same type were analyzed. Once the original had been approved, it was usually a simple matter for the group to analyze the appropriate matching assets, and five or six assets of the same type were sometimes analyzed in one day. As an example, all 180 Retail Lift Stations were completed in one month.

Few assets required less than six or more than ten meetings of 3 hours each. These meetings were usually scheduled at a rate of three meetings every two weeks; unless the facilitator's other commitments were too great. The meetings usually occurred as scheduled, with little reworking needed.

At first, three facilitators were trained by Broward County OES. One left for a job elsewhere, and two more were then trained. Thus at present, four trained facilitators are on the payroll. Each worked only one analysis at a time, and no more than three projects were in the meetings stage at any one time.

Altogether, 35 people were trained in the RCM2 process. All full review group members were drawn from this list. The list breaks down as shown below:

- 7 Electricians/Electrical Techs
- 6 Mechanics
- 3 Schedulers
- 2 Operators
- 5 Operations Supervisors
- 9 Maintenance Supervisors
- 2 Superintendents
- 1 Manager

An organization chart for Broward County is shown in Figure H.2.

Analyses were audited by a group of managers of the appropriate plant. This group consisted of the Superintendent, the Maintenance Supervisor, the Electrical Supervisor and the Chief Plant Operator of the appropriate plant. The audit meeting also involved all the group members who had worked on the analysis. The auditors were provided with copies of the draft one week before the meeting, and would ask for clarifications or make suggestions based on their experience. The most common source of alterations was the Superintendent, usually as a result of operational considerations that required a rescheduling of more complex PM's.

Outside consulting support was limited to facilitating the initial four analyses, training the personnel and periodic quality control reviews. Once the program was under way, Broward County employees were able to run the RCM project almost unaided. Broward County OES has appointed an RCM Program Leader – Ms Karen Valiquette who is a Chief Plant Operator, who became interested in the RCM program. Ms Valiquette assumed responsibility for managing the program – she still performs her CPO function in conjunction with her RCM duties.

Group members charged their time to the RCM project, leading to a total of $46,340.79 in labor costs. The total cost of the training supplied by the RCM2 consultants was $78,776. The total direct cost of the project was $125,116.79.

One Assistant Director of the EOD (Mr. Eugene Nicholas – the initiator of the program) was heavily involved in the process, meeting with the RCM program leader, Ms Karen Valiquette, regularly. Once a month, the program leader met with a group of executives to give a status report and answer questions.

Once an analysis had been approved by the auditors, the leader of the RCM program would make all changes to the Fixed Interval Maintenance work orders in the Computerized Maintenance Management System (CMMS) [Broward County OES utilizes MRO's MAXIMO software]. She would also rewrite operating procedures as needed. These changes were almost always made within a week of the analysis being given final approval. Substantial redesigns requiring the cooperation of the engineering department took longer, but few needed more than six months to begin work.

The Broward County RCM Program Leader is still in place, running the RCM program as part of her duties as a supervisor. She keeps track of maintenance activity through the Maximo system and meetings with maintenance personnel, and notes any significant failures that occur. She investigates the appropriate analysis, and would then reconvene the group if she feels substantial changes have to be made. No group has yet had to be reconvened. To date, on approximately six occasions, PM intervals have been changed as a result of experience. Several of these have been to lengthen intervals found to be unnecessarily conservative, and at least one was to undo a change made by management to the RCM recommendations.

### Overall Financial Benefits

The following table shows maintenance costs in dollars per thousand gallons for different parts of the EOD over the last seven years. The RCM program implementation began in earnest in 2002, so the average for the previous five years is compared to the average for 2002-2003 to show the effect of the RCM effort (Table 3.11).

To provide another perspective, the following information was extracted from the CMMS, using 2001 and 2003 data (Table 3.12).

**Table 3.11**
**Cost comparison in $/thousand gallons - before and after**

| Asset | 1997-2001 Average | 2002-2003 Average | Percent change |
|---|---|---|---|
| District 2A Drinking Water | 0.22 | 0.14 | -36% |
| Drinking Water Transmission | 0.28 | 0.02 | -93% |
| North Regional Well Field | 0.39 | 0.25 | -36% |
| District 1A Drinking Water | 0.36 | 0.29 | -19% |
| Sewer Collection System | 0.24 | 0.17 | -29% |
| Waste Water Treatment | 0.098 | 0.085 | -14% |
| Average | 0.265 | 0.159 | -40% |

**Table 3.12**
**CMMS cost comparison**

| Item | Before RCM (2001) | After RCM (2003) | Change in dollars |
|---|---|---|---|
| Emergency Work Orders | $209,567 | $143,016 | -$66,551 |
| Fixed-Interval Work Orders | $26,971 | $45,468 | +$18,497 |
| Total Expenditure | $236,538 | $188,484 | -$48,504 |

This works out to a 21% decrease in total maintenance costs for the entire facility. The CMMS is supposed to capture both labor and material costs but as with most non-integrated CMMS systems this does not always occur. The CMMS system reflects a rise in Fixed-Interval costs(which is in contradiction to the examples shown in Section 3) and a substantial decline in corrective maintenance costs. [The results from the financial systems shown above are the more accurate figures in terms of maintenance cost savings]. The reason for the increase in fixed-interval costs can be attributed to the inclusion of tasks to manage hidden failures (failure-finding tasks/function checks); development of optimized maintenance programs for certain assets which did not have any programs previously and to more intensive use of the CMMS system and improved real completion rates on fixed-interval tasks once the RCM initiative developed a greater sense of ownership and commitment.

These two tables clearly show that the Broward County EOD derived a HIGHLY significant financial benefit from the application of RCM. The second table shows a sharp fall in repair costs, illustrating that the source of the cost saving was an increase in reliability.

## Intangible Benefits of RCM

### *Operations-Maintenance Communication*

For several years prior to the RCM program, EOD management had been trying to promote team-building concepts across all three facilities, but especially between Operations and Maintenance. The RCM2 process involves people from all ranks of both groups sitting down together to discuss the equipment in detail. Often, this is the first time the two sides have met to talk without their being any question of blame for some recent incident. This was the case to a significant degree in the EOD as well. Consequently, communication between the two departments was improved by a large amount, although obviously it is difficult to quantify this gain. Also, the regular meetings between managers and supervisors resulted, in the opinion of both sets of leaders, with improved communications.

Pat Cerbone, a Chief Plant Operator, supplied the following quotes about this phenomenon:

> "The ability of maintenance and operations personnel to work cohesively when facing problems from a common perspective was a revelation."

> "Working on these RCM analyses, and the decision making process's involved, has given me a better understanding of managements problems regarding resource allocation."

### *Knowledge of Asset and Process Function*

The extensive application of RCM II resulted in a gain of knowledge about the assets by the personnel involved that spread throughout the organization through altered training and operating procedures. This again is difficult to quantify, but management and supervisors both mentioned it readily and easily. Regular, blue-collar employees felt much more involved in the process of maintaining the asset systems after having played a key part in formulating the maintenance plan utilizing their experience and knowledge with these assets. This led to a considerable increase, according to all available evidence, in the PM completion rate

Metason Phillip, a Maintenance Supervisor, had the following quote:

> "Participating in the RCM program has deepened my knowledge of preventive maintenance and how it should be performed, and increased my understanding of the importance that management and craft must play in order to have a successful RCM program. Involvement of management as the driving force to both complement and support the EOD's craftsmen to be willing participants in terms of ownership and responsibility of his/her equipment has enabled the achievement of excellence in maintenance as a unit- that is what RCM is all about."

### *Knowledge Retention*

The EOD, like so much of industry, has a skilled workforce that is disproportionately over the age of forty. Consequently, a great loss of accumulated experience will occur over the next 10-15 years. The RCM program collected the wisdom of dozens of experienced people in one place, and provided the inspiration for numerous amendments to training procedures. This will help mitigate this problem.

### *CAD Drawings and Documentation Update*

Finally, the RCM process involves workers intimately familiar with a given asset working through it component by component. As a result, any errors on the associated drawings will be discovered, and therefore more accurate documents will follow from the meetings.

### Conclusions On Broward County RCM Program

The RCM II program at the Broward EOD was motivated predominantly by the desire to control maintenance costs. The plants involved did not have significant environmental or safety problems. Equipment availability was a secondary consideration. The above data shows clearly that both the primary goal of reducing costs and the secondary goal of increasing availability by reducing unscheduled outages were accomplished. The unplanned intangible gains were also noted, and well appreciated and will continue to contribute to the successful accomplishment of the primary goals.

> "Our division achieved excellent results associated with implementing a RCM program such as PM & PdM tasks, failure modes, functional failure, etc. and the financial benefit derived from implementing these tasks. In addition our Division achieved new levels of understanding and cooperation between operations and maintenance personnel as a result of them working together on the RCM teams"
>
> *Gene Nicholas, Associate Director, Environmental Operations Division*

[Detail is included in Appendix H]

## SUMMARY OF MWRA EVALUATION - THE DITP RCM PROGRAM

The MWRA's Deer Island Treatment Plant became operational in the mid 1990s. It replaced a plant that had been decommissioned early, largely due to poor maintenance practices. As a result, DITP management was committed to establishing a world-class maintenance program for the new plant. This lead to the Facilities Asset Management Program, or FAMP. The MWRA solicited bids from several consulting firms to assist with the planning and execution of the Facilities Asset Management Program, and selected New Dimension Solutions (formerly Spearhead System Consultants) – a consulting firm based in New York which specializes in driving value from best practice Enterprise Asset Management and Reliability.

### Facilities Asset Management Program

Upon taking over responsibilities as the Deputy Director, Maintenance, of the Deer Island Facility in 1996, Gerry Gallinaro initiated the Facilities Asset Management Program (Figure 3.3). Staff researched various programs utilized throughout industry and elected to solicit assistance from Engineering firms to provide final review and piloting of the selected program. During the initial review process the MWRA decided to expand the Program to encompass the entire Operations including the wastewater collections and water treatment/distribution. The Facilities Asset Management Program can be further described as having two objectives:

- To cost-effectively replace the less durable capital components of the facilities at the appropriate time to ensure reliable plant operation and preserve the value of the original investment.
- To prolong the equipment life and control the rate of replacement of major asset systems (i.e. to avoid large spending spikes for consolidated retrofit or rehabilitation).

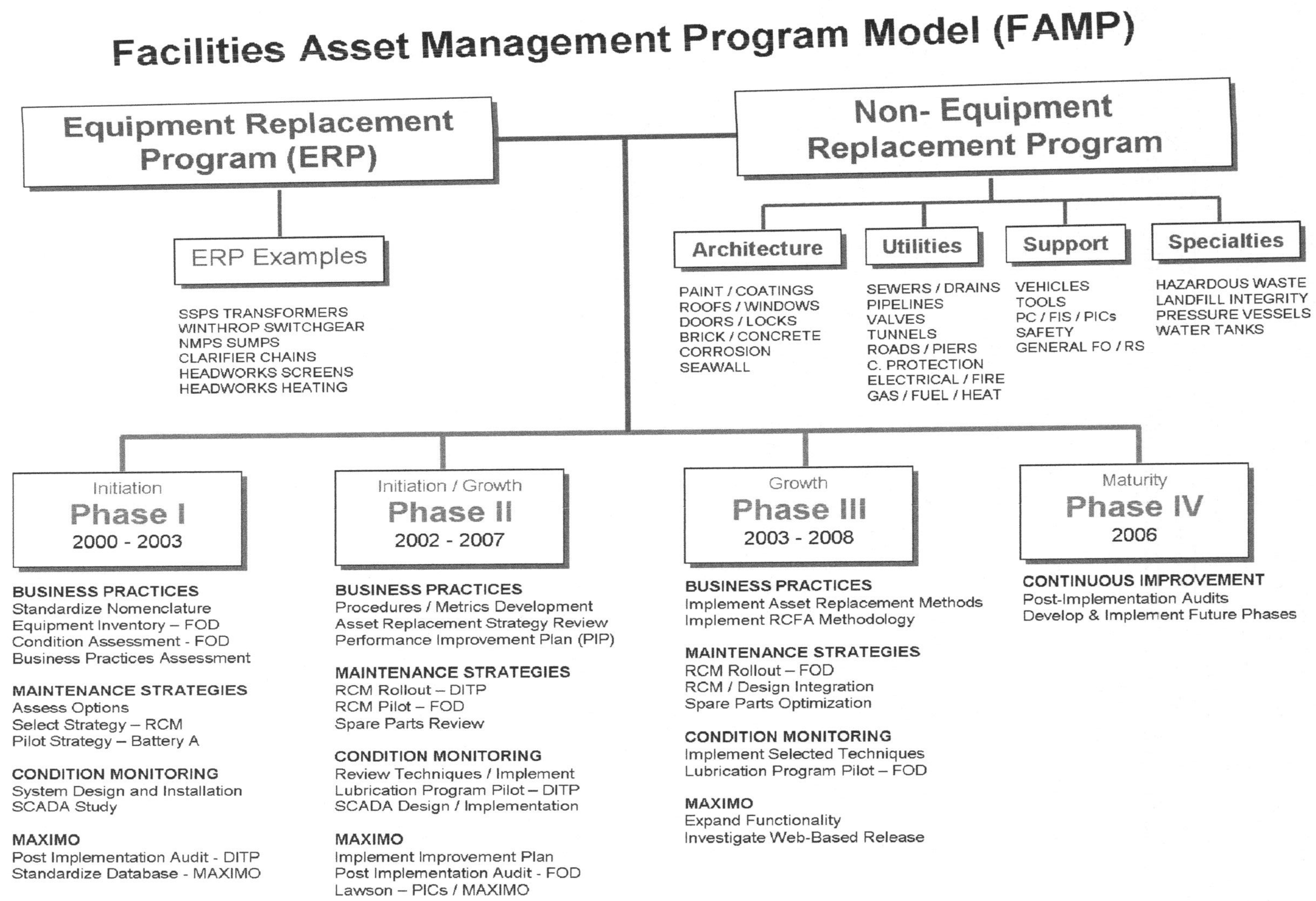


**Figure 3.3 MWRA's Facilities Asset Management Program**

The Authority issued a Request for Qualifications/Proposal (RFQ/P) to secure professional services to develop the most efficient strategy to integrate maintenance, operations, and engineering activities to support the FAMP objectives in October 1999. NDS responded to the RFQ/P and was awarded the contract in March 2000.

In order to implement the FAMP objectives, NDS was required to perform several tasks. These are detailed in Attachment D of Appendix I and included standardization of the maintenance management systems, practices and procedures for the Operations Division. As noted above, during the initial review process for FAMP, the MWRA decided to expand the Program to encompass the entire Operations including the wastewater collections and water treatment/distribution. Additionally NDS evaluated various maintenance strategy development practices. This study selected RCM as the optimal strategy for the MWRA.

**RCM Pilot Program on DITP's Primary Clarifier Battery A**

RCM was implemented on DITP's Primary Clarifier Battery A as a pilot project. Twelve analyses were completed on Primary Clarifier Battery A, which resulted in significant cost savings. These analyses were then used as templates for the remaining Primary Clarifiers. RCM2 has now been adopted as MWRA's maintenance strategy development tool of choice and has also been rolled out to Field Operations Division (FOD) (Waterworks and Transport divisions). NDS was awarded an additional contract to provide additional RCM services to extend the RCM program to additional sections of the DITP facility and other areas of the MWRA.

The pilot program consisted of twelve systems, started on the dates shown, all from Primary Battery A, or A and B combined:

| | |
|---|---|
| • Primary Sludge Pumps | July 2000 |
| • Primary Scum Removal System | August 2000 |
| • Channel Aeration Blowers | October 2000 |
| • Primary Sludge Longitudinal Collectors | November 2000 |
| • Primary Gallery A/B HVAC System | December 2000 |
| • Influent and Effluent Sample Pumps | January 2001 |
| • Chlorine Gas Detector System | February 2001 |
| • Hot Water (W5) Flush System | March 2001 |
| • EB1 Power Supply | January 2001 |
| • A/B Building Fire Systems | March 2001 |
| • Primary Sludge Cross Collectors | March 2001 |
| • Primary Sump Pumps | April 2001 |

**RCM Program Roll-out**

After the pilots were completed, the RCM program was extended and rolled out to replicate the above for the other three batteries and extend on to other assets as listed below:

| | |
|---|---|
| • B/C/D Influent and Effluent Sample Pumps | August 2001 |
| • C/D Building Fire Systems | September 2001 |
| • B/C/D Chlorine Gas Detectors | October 2001 |
| • B/C/D Cross Collectors | October 2001 |

- B/C/D Primary Sludge Pumps December 2001
- B/C/D Primary Scum Removal System January 2002
- C/D Primary Gallery HVAC System January 2002
- B/C/D Primary Sump Pumps November 2001
- EB2/EB3 Power Supply May 2002
- RSL Pumps July 2002
- East Odor Control Scrubber July 2002
- Main Cryogenics Air Compressors September 2002
- Digester Gas Compressors September 2002
- North Main Pumps November 2002
- A Hypochlorite Feed Pumps December 2002
- South Main Pumps February 2002
- Sodium Bisulfate Pumps - Dechlorination January 2003
- W3L Pumps March 2003
- WSL Centrifuges April 2003
- Battery A/B Secondary Sludge Collectors July 2003
- Channel Aeration Review July 2003
- NMPS HVAC August 2003
- Battery A/B Secondary Scum System August 2003
- Winthrop HVAC System September 2003
- Winthrop Pumps September 2003
- NMPS Hydraulics October 2003
- Hypochlorite Receiving September 2003
- Low Load Cryogenic Air Compressors November 2003
- Digester Flares December 2003

These assets were selected using the criticality matrix developed by MWRA staff with the assistance of NDS. NDS developed an MS-ACCESS database to input the criticality results and print reports. MWRA staff reviewed each system in all the MWRA facilities (DITP and Field Operations), noted its consequence and likelihood scores, and then multiplied them together. These products were ranked, and the most critical selected for analysis. After the pilot, the MWRA continued to work down the list. At the present time, this process is still ongoing, at a rate of about twenty analyses per year. Management estimates about 100 more systems will be analyzed in the next five years.

The process used to perform and implement RCM analyses at the MWRA is shown in Figure 3.4.

**Notes and Comments on RCM Program**

As mentioned above, DITP is divided into four primary batteries, each with very similar equipment. Therefore, equipment in Battery A was selected to be analyzed, and then the Primary Battery A analyses were used as ***templates*** and adapted for Primary Batteries B, C and D.

In most cases, there were no major design differences between assets in each Primary Battery and the base-line. In several cases, one primary battery had had some additional ancillary equipment installed, but this seldom required more than one meeting to analyze.

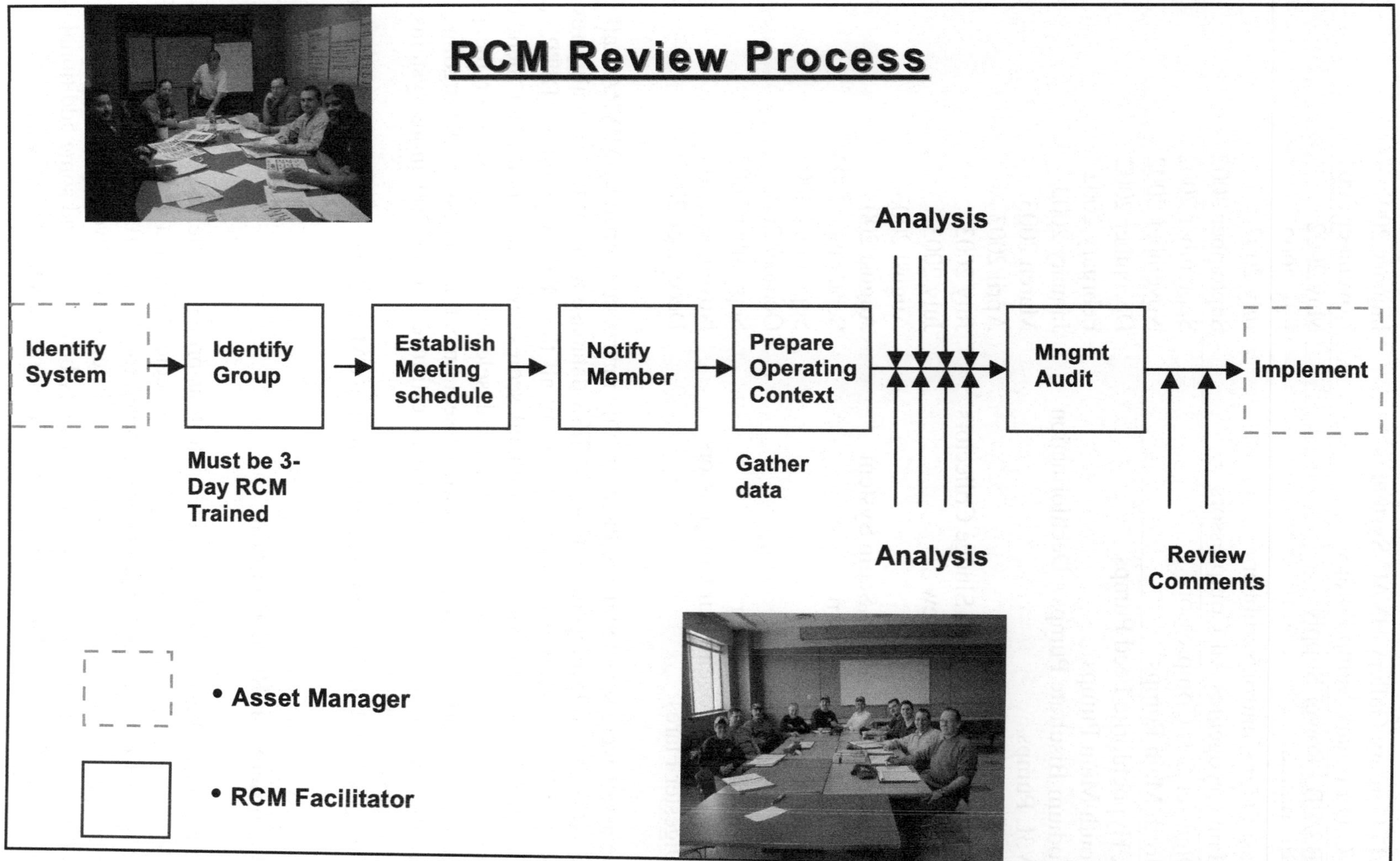


Figure 3.4 RCM review process flow implemented by MWRA for FAMP program

After the audit team had signed off on a new analysis, the group would reconvene with all appropriate drawings and work through the analysis to make the necessary changes for each replication/variant analysis using the initial asset system analysis as a template. This usually took one meeting of three hours, in which time several replications or variants could be made.

The database shows that, in most cases, one analysis was completed and audited, a process that usually took about two months, before the replications/variants were created. Once an original was approved, a period of between three and twelve months elapsed before the replicant/variant analyses were started. The time required to take a replicant/variant analysis to the final sign off point ranged from one to fifteen months.

Few analyses of initial asset systems required less than six or more than ten meetings of 4 hours each. For the pilot projects, meetings were scheduled daily, from 7:30 to 11:30. For the later analyses, meetings were scheduled every other day at the same hours.

A total of twenty three MWRA employees were trained as facilitators as part of the FAMP program. Of these, twelve are associated with DITP, while the other nine work for the Field Operations Division.

Altogether, 290 people from DITP and FOD have been trained to date in the RCM II process. All full review group members were drawn from this list.

Analyses were audited by a group of managers of the appropriate plant. The DITP group consisted of the Director; Deputy Director, Maintenance; Deputy Director, Operations; Manager of Capital Engineering, Assets Manager and Work Co-ordination Manager. All members of the review group were also invited, and about fifty percent attended. The auditors were provided with copies of the draft one week before the meeting, and would ask for clarifications or make suggestions based on their experience. A typical audit analysis produced between ten and twenty changes, most of them minor rewording of failure effect descriptions.

NDS provided all the consulting services to the RCM program. This involved training all the above-mentioned personnel, facilitating a number of the initial analyses, reviewing the facilitator analyses, and coaching the MWRA facilitators until they reached a high level of skill.

The group members involved in the RCM program had to divert a certain amount of time from their regular jobs. This was somewhat disruptive to the normal work schedule, but management was able to limit problems by choosing times and participants carefully.

The cost of NDS' services directly related to RCM at Deer Island has been $496,969.32 to date, over a period of several years.

DITP management formed a total of ten task teams to keep control of the various components of the FAMP program. One of these is the RCM group. This group meets regularly, and posts a progress update on the corporate web site every month.

The person most responsible for running the RCM program at DITP is John Colbert, the Asset Manager. He leads the task team. This group works with the appropriate division managers to select members of new analysis teams, and to develop the schedule of analyses.

After an analysis has been signed off by the audit team, the Operations department takes the list of changes to procedures and implements them. The Maintenance department makes all approved PM changes. John Colbert has developed a spreadsheet on which he keeps track of all these, as well as all the one-time adjustments developed by the analyses. It is his job to ensure that the Operations and Maintenance departments remain on track, and that redesigns or training talks occur within a reasonable time.

**Table 3.13**
**Comparison of preventive and corrective hours before and after RCM**

| | 2000 | 2003 | Change |
|---|---|---|---|
| PM Work Orders | 40,000 | 24,000 | -16,000 |
| CM Work Orders | 6,800 | 4,100 | -2,700 |
| CM/Project Hours | 120,000 | 100,000 | -20,000 |
| PM Hours | 80,000 | 55,000 | -25,000 |

[PM = Fixed interval work orders; CM = Corrective maintenance work orders]

## Overall Financial Effect of the RCM Program

The best illustration of the overall effect of the RCM2 program on DITP maintenance costs and practices can be seen in the attached charts (Attachment A – 1998-2003 and Attachment B YTD 2004 in Appendix I of this report). Some of the data from these charts is summarized in Table 3.13.

All the RCM2 analyses done at DITP included a summary of the predicted benefits. In the later analyses, this incorporated a comparison of the existing PM regime with that recommended by RCM2. On average, this showed a drop of about 130 hours per year per analysis. For 33 completed analyses, this works out to 4290 hours per year, which works out to an annual saving of at least $150,000 from RCM. The rest of the decline in PM hours is the result of other PM consolidation efforts, many also closely involving NDS.

The decline in CM of 20,000 hours works out to a saving of $700,000 per year. How much of this is due to RCM is not possible to quantify. The reduction in CM hours does show that the Fixed Interval (PM) reduction certainly did not result in additional corrective maintenance work. It is indeed apparent that the effective maintenance applied as a result of the RCM analyses, has supported a substantial reduction in corrective maintenance work.

In summary over the three year period from 2000 to 2003, as results of the FAMP project took effect, the hours required for Fixed-Interval PM work reduced from 80,000 hours per year down to 55,000. During the same period, Corrective Maintenance and project hours also declined from 120,000 to 100,000 hours. This means that maintenance hours in total declined by 45,000 hours per year during this period (c 25%). Plant availability was consistently maintained in the upper nineties (%) during the same period.

One of the motivators for MWRA management regarding the FAMP project was the concern that maintenance costs would start to rise as the plant aged (DITP was commissioned in the mid nineties). The FAMP project, using RCM as its foundation stone, has instead accomplished a major increase in maintenance productivity.

A subset of maintenance KPI's from the MWRA over the period 1998 to 2003 is shown in Figure 3.5. This shows the impact of the FAMP program and its RCM-based Reliability Initiative on performance of the maintenance function. As show in Appendix I other key asset performance metrics such as Availability also consistently improved over this period, although the cost and amount of maintenance was significantly reduced.

# Yearly Maintenance Metrics
## Overall Maintenance Program Measures

### Overtime Percentage

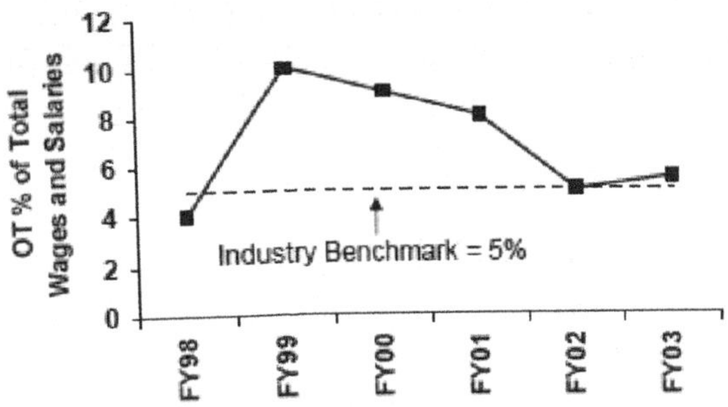


Overtime has decreased steadily from FY99 to FY02. The slight increase in FY03 is attributed to a loss of staff to early retirements and internal transfers which increased maintenance backlog. Additional OT expenditures into FY04 will be required until existing vacancies are backfilled.

### Craft Work Orders

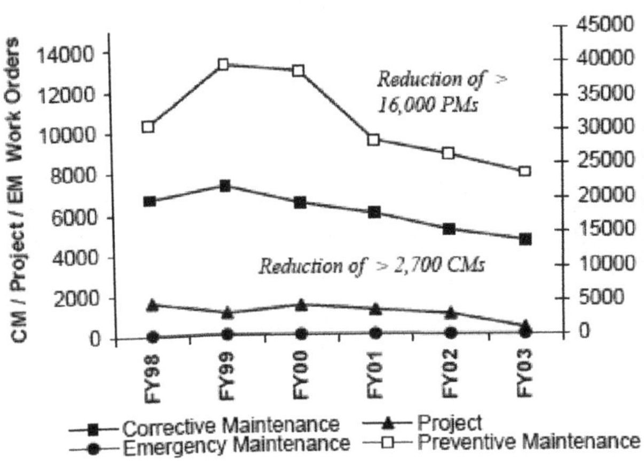


The number of PMs has been significantly decreased due to a concerted effort to optimize the PM program based upon staff input, elimination of duplicate PMs, and implementing Reliability Centered Maintenance recommendations. The number of CMs and Project work orders have decreased as startup issues and initial design issues are resolved. The number of EM work orders continues to be very low showing that staff continue to maintain the appropriate level of equipment to support the NPDES permit. In upcoming years as the plant ages, it is expected that a leveling out and eventually an increase in CM and project work will occur.

### Craft Hours Distribution

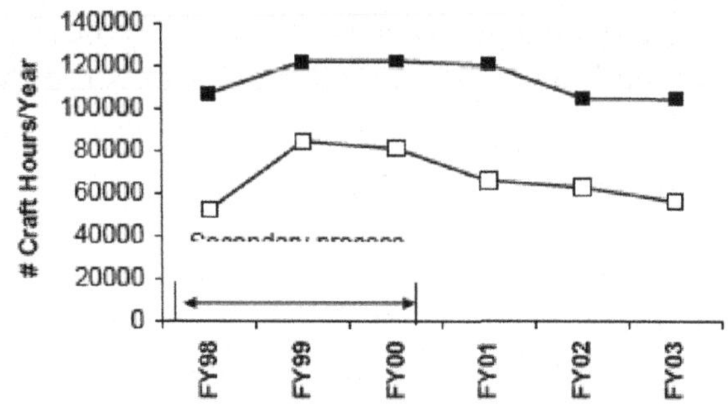


Optimization of the preventive maintenance program through the transfer of some light maintenance to Operations (11% of the PM hours at the end of FY03), elimination of duplicate work orders, decreasing PM frequency due to equipment history and performance, and Reliability Centered Maintenance recommendations has resulted in a significant decrease (28,300) of PM craft hours from FY99 to FY03

**Figure 3.5 MWRA KPI trending chart 2**

## Intangible Benefits

### *Knowledge of the Assets*

The greatest intangible gain from the RCM2 program in the opinion of DITP management was knowledge of the assets on the part of operators and maintainers. Working through the details of each system in what is still a fairly new plant, resulted in all group members learning much about the inner workings of the systems they work with, as well as a clearer view of how each fits into the overall plant design and process flow. These gains are illustrated by the following quotations from DITP employees:

> "I thought I knew everything about this system, but after a few RCM analysis sessions, I realized I only knew about 10%. Now, through the knowledge transfer that occurred during RCM, I could teach a course."
>
> *Operator's comment*

> "This is more than I ever wanted to know about this system."
>
> *Mechanic's comment*

As is implied in the first quotation, all the knowledge gathered from the different members of the review groups is kept within the utility to a much greater degree than before, reducing the impact of the staff losses DITP is currently experiencing.

### *Communication and Teamwork*

In the opinion of DITP management, the next most significant intangible effect of the RCM program was the improvement it generated in communication, both between Operations and Maintenance and between management and the workforce. Over the last couple of years, DITP has reduced its workforce considerably, and this generated some expected suspicion from the workers about the real intent behind the RCM2 initiative. This was eventually dispelled, as shown in the following quotation:

> "I was very skeptical. I thought this would be another tool for cutting the workforce. Now I am impressed with the level of detail that RCM actually goes into. The people that actually do the work are the ones making the decisions on how the equipment is maintained."
>
> *HVAC Technician's comment*

### *Ownership of the Maintenance Program*

One effect of the involvement of the maintenance technicians' input in the development of maintenance strategies that the RCM2 process encourages is a greater feeling of ownership on the part of the technicians who actually perform the maintenance. The best demonstration of this is, as can be seen in the attached charts, the fact that the PM completion rate has risen from 82% to 92% since the RCM program began in 2001.

"I'll do more analyses. I'm sold. I think we're creating something good here."

*Mechanic's comment*

### *Drawings*

As DITP is a relatively new plant, a reasonably complete and current set of drawings was in place before the RCM analyses began. However, the close attention paid to the drawings by the analysis teams often revealed a number of small changes to be made to bring drawings up-to-date with the real-world configuration.

### *Safety Benefits*

Safety was not a significant problem at DITP before the RCM2 program began. However, a number of RCM2 analyses have produced significant recommendations to improve safety. Many of these recommendations have been simply to add extra warning signs or lights in areas that may have been overlooked by the initial builders of the site. Other recommendations have reworked some operating procedures to eliminate unnecessary risks, or led to the replacement of outdated or obsolete safety equipment, such as an old obsolete Chlorine monitor in one Primary Battery.

The most interesting safety enhancement resulting from the RCM2 initiative resulted from the decision to apply the RCM Review Group approach to revising a confined-space work procedure that had caused considerable friction between management and maintenance workers. This led to the creation of a new procedure that ended the often sharp debate, and was completely acceptable to all sides. This also contributed to the improvement in corporate communications created by RCM2.

## Conclusions on the DITP RCM2 Program

The RCM2 program was launched with two main purposes in mind. The first was to prolong the equipment life to ensure the plant will remain in service for at least its intended life of 50 years and the second to control the rate of replacement of major asset systems (See Section 3.1). The use of RCM for the first objective cannot be definitively determined now, but the decline in Corrective Maintenance observed since the beginning of the initiative bodes well for the future. The use of the "duty-standby" methodology for many assets as a result of the RCM analysis will help meeting the second objective. All assets will not have equal wear and not all require replacement at the same time.

An additional benefit is that the optimization of the site's maintenance program (fixed interval tasks) and the reworking of operating procedures resulting from the RCM initiative have resulted in significant cost savings by reducing labor hours.

[Details are included in Appendix I]

## UTILITIES SURVEY ON MAINTENANCE PROGRAM DEVELOPMENT AND RCM

A substantial survey was effort was undertaken among water utilities in North America. A questionnaire was carefully designed to support the aims of the research project and was set up as an on-line survey. NDS personnel contacted a large number of utilities and ultimately secured a response from approximately 200 utilities. The results of the utilities survey are included as Appendix B.

## EXECUTIVE SUMMARY OF UTILITIES SURVEY

A total of 177 respondents representing water utilities across the United States and in Canada completed an online survey that was developed by New Dimension Solutions. The survey instrument was designed to establish personal data of the respondent, develop a profile of the utility that the respondent represents, determine maintenance programs and efforts currently made within their facilities, and ascertain their involvement with and any benefits achieved through the application of Reliability-Centered Maintenance (RCM).

The survey participants included 177 respondents with a representative mix of utility sizes for <10 MGD to >300 MGD. The majority of the respondents were city- or municipality-owned (77%). The majority performed their own maintenance (>85%).

### Current Maintenance / Asset Management Performance and Priority

Data revealed that there is an emergent focus by utility management on asset performance and optimal management of the utility's assets. Over half of respondents (57%) indicated that this focus is becoming an important issue for utility management, and 13% of respondents stated that optimal management and performance of the utility's assets is becoming ***a very critical issue for utility management***. Respondents were also asked to provide reasons for any change in management prioritization of physical asset maintenance and performance. When the first two reasons for this change in management prioritization were grouped together, aging equipment, facilities, or infrastructure (21%), the desire for better performance, higher efficiency, or optimization (16%), the need for cost reduction or to control costs (10%), and for compliance with regulatory requirements or more stringent regulations (10%) were the most commonly mentioned responses.

Respondents were also asked to specify the most serious issues that currently face utility management, in terms of the maintenance of its assets. When the top three most serious issues were grouped together, capital replacement costs (45% included this in their top three) was the most commonly mentioned, followed by maintenance costs (44%), aging facilities (42%), and reliability (40%).

To gain an understanding of the manner in which maintenance departments function across the various utilities, respondents were asked to select the mode of operation of their maintenance department. A majority of respondents (61%) indicated that their maintenance department function operates in a steadily improving proactive mode, recognizing that some asset systems and facilities continue to be a problem. Twenty-two percent (22%) of respondents say that their maintenance department function operates with the basics of a proactive program, but that maintenance is still mostly reactive in nature.

Plant breakdowns are defined in the survey as being unscheduled failures of an asset that affect plant operations and must be rectified immediately when the asset fails. This is an extremely important issue for utilities with highly loaded or critical assets. According to the survey, on average, 20% of maintenance work is in direct response to these plant breakdowns.

Managing assets via a Computer Maintenance Management System (CMMS) or an Enterprise Asset Management (EAM) system involves tagging physical assets and registration of these assets within the system. The degree to which utilities tagged physical assets and registered these assets within the CMMS / EAM system differed across respondents. For example, 35% of respondents said that the tagging and registering of some critical assets have been completed, yet there was still much more work to be done. Thirty-one percent (31%) indicated that the tagging and registering of most assets has been completed, including all critical assets, with some clean-up work still to be done. ***Another 31% said that no assets at all have been tagged or registered.***

Fixed-interval maintenance tasks (also known as preventive maintenance tasks) are developed from several methods including OEM recommendations, equipment history, formalized preventive maintenance optimization processes, and RCM. When asked to specify the percentage that each of these methods have contributed to the development of the fixed-interval tasks at their utility, respondents on average assigned 48% to OEM recommendations, 31% to equipment history, 10% to formalized preventive maintenance optimization process, and 3% to RCM.

## Asset Management Programs – Key Points From Survey

Asset management is becoming an important issue with 70% of water utilities. The majority of water utilities believe there is room for improvement in their current utility asset management program and that they are steadily working to improve their programs. Senior management commitment to asset management was developing or strong for greater than two thirds of respondents.

The asset management program for the majority of utilities is in the starting or initial phases with another 22% with no asset management program. The majority of critical assets were included in the asset management programs. The utilities considered that the overall asset condition was the same or improving at over 75% of all utilities with 22% stating that asset condition was deteriorating.

The top three reasons for the formation of an asset management program are:

- Aging infrastructure
- Better performance
- Cost-savings/Reductions

## Maintenance Programs / Failure Management Policies

Formal and systematic processes and methods to derive maintenance programs generally appear to be in place in many facilities. Forty-five percent (45%) of respondents indicated that these methods have been deployed or utilized at their utility on certain critical assets, and 14% said that a comprehensive program utilizing formal processes to derive asset maintenance programs is in place across the utility. Another quarter (26%) stated that there is a sporadic effort to implement formal and systematic processes / methods at their utility, and 15% said that there are no formal or systematic processes / methods in place at the utility to facilitate the development of maintenance programs for the utility's assets.

Respondents were asked to indicate their level of familiarity with the main Maintenance Strategy/Program Formulation Techniques currently available for developing optimized maintenance programs for physical assets. Familiarity with Original Equipment Manufacturer (OEM) Programs was highest (74%) followed by PM Optimization (46%), RCM (36%), and Root Cause Analysis (34%). In addition to these, respondents were asked to specify which maintenance strategy formulation techniques were used to develop maintenance programs for critical and non-critical physical assets. In both of these cases, the Original Equipment Manufacturer Program was used on the highest percentage of physical assets, followed by PM Optimization, and then RCM.

Applying formalized maintenance strategy formulation techniques or reliability optimization methodologies during the design-build phase of new assets can optimize future maintenance of assets and minimize life-cycle costs. Respondents were asked whether such Maintenance Strategy/program Formulation or Reliability Optimization techniques are applied at their utilities either during the design-build phase of new assets, or during the modification of existing assets. Approximately 44% of respondents indicated that the maintenance staff performs some formalized reliability review of the proposed design, and 31% indicated that design engineers and draughtsman try to consider reliability and maintainability during design. About 22% stated that no formalized maintenance strategy formulation techniques or reliability optimization methodologies are applied in the design-build phase or when modifying existing assets.

With respect to formalized Criticality Analyses for the technical systems or assets within the utility facilities, 44% of respondents indicated that no such analyses have been performed at their utility. One third of respondents (32%) stated that this type of analysis has been performed for some areas of the plant, and one fifth (20%) said that a formalized Criticality Analysis has been performed for most areas of most facilities.

An effective reliability program includes activities such as regular reviews and adjustments to the maintenance program and standard operating procedures. Nearly half (47%) of applicable respondents indicated that some asset failures and performance shortfalls are reviewed to identify cause and to then develop appropriate failure management policies, and 29% indicated that asset failures and performance shortfalls are reviewed for critical equipment.

Reviews of asset maintenance programs appear to be a common task among respondents, as over half (52%) indicated that some of these programs are reviewed on a regular basis. About 22% indicated that these programs are reviewed on a regular basis for critical equipment, and 12% said that all of these programs are reviewed on a regular basis. When maintenance programs are developed, over half (52%) of applicable respondents stated that some of the process steps used to develop the maintenance programs are formally documented. Nearly one quarter (24%) said that none of these steps are documented.

Earlier, it was noted that reliability is among the top three most serious asset maintenance issues facing utility management. Given the severity of this issue it is no surprise that there is an emergent awareness among senior management with respect to their commitment to reliability initiatives. More specifically, 39% of respondents feel they have strong senior management commitment towards reliability improvement initiatives, and 29% feel management is aware of these initiatives. When asked whether or not their utility has any formalized strategic programs in place to improve reliability of the utility's physical assets, 30% of respondents indicated that their utility is starting a formal program to improve asset reliability, and one quarter (25%) said that there are initial discussions and plans to commence such a program.

Approximately three quarters (74%) of respondents said that their utilities do not have a reliability group in place at their utility. One fifth (21%) of respondents said that an engineer on a part-time basis fulfills the reliability group function.

Utilities generally appear to be making an attempt at keeping up-to-date with new initiatives and industry standards. More specifically, 37% of respondents indicated that their utility takes part in some benchmarking and networking activities and 27% said that their utility belongs to organizations.

## Reliability-Centered Maintenance (RCM)

Of the 177 valid utility responses, 42 went through the section on RCM. [Participation in this section was limited to utilities "where RCM has been used in a significant manner"].

### *Key Points on RCM From Utility Survey*

The majority of water utilities use OEM recommendations to formulate maintenance strategy for critical and non-critical assets. RCM is not yet widely used by the water utilities industry (only 10% of the 177 respondents responded to the majority of the RCM questions.)

83% of 18 utilities recognized a positive payback on the RCM effort, with five utilities stating the payback was within 3 months to two years.

The major benefits from the RCM program were stated to be:

- Knowledge Retention
- Asset and System Knowledge
- Improved maintenance work completion
- Reduced maintenance cost
- Operations and Maintenance Co-operation

### *Other Salient Data Points on RCM in Utility Survey*

Respondents from those utilities that use RCM were asked to indicate the extent to which RCM has benefited the utility. Knowledge retention for the utility was the most commonly mentioned benefit (60%). Knowledge of assets / systems was another commonly mentioned benefit (58%) along with improved completion rate on maintenance tasks, and reduced maintenance cost (54%). Regarding reliability, 16 % saw significant improvement and 32% saw improvement in some cases.

Benefits were surveyed for thirteen variables. The sample of respondents on benefits ranged from 23 to 25. Across the thirteen variables surveyed, the majority of respondents saw benefit raging from marginal to significant, with approximately one-third seeing no benefit.

In terms of their importance to the utility, respondents ranked reliability as the most important benefit, followed by cost reduction and then knowledge retention.

Derivation and calculation of required frequencies for failure finding tasks for hidden failures is performed in some RCM methodologies (32%). Fully 40% weren't sure whether or not the RCM Methodology used includes this type of calculation.

Respondents were also asked to state their level of awareness with SAE Standard JA1011, which defines standards for acceptable RCM processes. Among applicable respondents, only 12% are familiar with it, and 14% have heard of it, but are not familiar with the details.

## Potential Benefits From Use of RCM Based on Survey Results

### *Duty-Standby Strategies*

Utilities capacity was less than demand for 97% of the utilities surveyed (i.e. the utilities had some excess capacity). The RCM process uses duty-standby strategies to extend equipment life and improve asset reliability. The excess capacity of utilities in normal load conditions, can be leveraged to provide maintenance cost savings and improve equipment reliability by using optimized duty-standby strategies.

### *Reduction of Emergency/Breakdown Work*

The failure of assets that affect plant operations that must be rectified immediately (emergency work) exceeds 10% of all work at the majority of utilities. The industry benchmark for emergency work is <5%. The use of RCM can significantly improve equipment availability and assist in improving plant performance by the use of standby-duty operation, improving the preventive maintenance of asset components by the use of condition monitoring techniques and the elimination of counter-productive, intrusive tasks and identifying appropriate maintenance tasks/functional checks for hidden failures.

### *OEM-Recommended Maintenance Programs Implies Opportunity for Optimization*

The preventive maintenance programs for approximately half of the utilities surveyed are based on OEM recommendations. The use of OEM-based maintenance programs does not account for the operating context of the equipment in use. The operating context of each plant's equipment (assets) is reviewed during the RCM analysis process and considers the actual operating conditions such as flow, pressure, temperature and process fluid used by each component and its effect on the optimal maintenance required by that asset in those conditions. For example, a sump pump that operates infrequently and is backed up by standby unit may be run to failure, while a critical raw waste water pump that runs 24/7 may require extensive condition monitoring to optimize its availability and avoid consequences of unexpected failure. Plainly the OEM recommendations cannot fit every operating context – even if they have been compiled based on Reliability best practice and RCM concepts (which is frequently not the case). One size does not fit all. Thus this finding that over 50% of maintenance programs are based on OEM recommendations implies that there is much scope for optimization of maintenance programs using RCM.

### *Lean Optimized PM Program Leading to High Completion Rates*

The survey results indicate that the preventive maintenance tasks at water utilities are scheduled approximately half the time (48%) and 59% of these tasks are completed. The correct scheduling and completion of preventive maintenance tasks is important to ensure equipment reliability and availability. The establishment and completion of "PMs" is one of the basic building blocks of an asset management program. The industry benchmark for completion of PM tasks is 90%. RCM can improve the preventive maintenance program performance by eliminating low-value PMs, correcting the frequencies on PMs which are set or based on incorrect premise and ensuring that only required PMs based on sound concepts remain. Implementing RCM based on the recommendations laid out in this survey also esnrues that maintenance and operations personnel associated with the asset take ownership of the PM program. Thus RCM programs lead to high completion rates by leaning the PM program to an optimized set of effective tasks while simultaneously building ownership of the program with the people that do the PM tasks and operate the assets.

## Other Asset Management Issues

### *Asset Tagging*

The first step in any asset management program is to identify all assets to be managed/maintained at the utility. Asset tagging using a unique identifier is required to do this to link information to and track activity against that asset. Two thirds of respondents stated that some critical assets were tagged and that much work needed to be done to complete tagging.

### *CMMS System*

Implementation of a computerized maintenance management system (CMMS) is often the next step in the development of an asset management program. CMMS are often not used for small utilities due to the lesser number of assets and size of maintenance staff. The majority of utilities (54%) use a CMMS. Those using CMMS include > 70% of assets in the CMMS database. The average use of the CMMS was five years.

### *Asset Criticality Rating*

The criticality of assets was not established for approximately half of all utilities (44%). The establishment of asset criticality is an important step in work prioritization, budgeting and implementation of an asset management program. Detail is included in Appendix B.

## LITERATURE SURVEY

A literature survey was undertaken with intensive on-line research. Results of this survey are included as Appendix A.

# CHAPTER 4
# SUMMARY AND CONCLUSIONS

## OBJECTIVE OF THE PROJECT

The objective of the project is to "assess how utilities can apply RCM to new and existing infrastructure and evaluate the costs and benefits of such a program."

## HOW UTILITIES CAN APPLY RCM TO NEW AND EXISTING INFRASTRUCTURE

Chapter 5 provides detailed recommendations on exactly how utilities should go about implementing RCM to both new and existing infrastructure.

These recommendations are validated by the Pilot Sites performed as part of this project at Denver Water and at Veolia Water International. (Appendixes F and G) and the detailed evaluations of established successful Reliability programs based on RCM at Broward County in Florida and MWRA in Massachusetts (Appendixes H and I). It is interesting to note that Denver Water has embarked on a strategic Reliability Program using RCM based on the success of the Pilot project. Veolia Water Indianapolis has also made RCM a cornerstone of its innovative partnership with the City of Indianapolis – see excerpt from 2004 Annual Report in Figure 4.1.

The pilots and evaluations focus primarily on existing assets. With respect to new assets the RCM program provides greater benefit as the design process can be optimized for improved reliability and maintainability without the constraints and costs of modifying existing real assets. The authors of the report are currently performing RCM work on a major new pump station for Orange County Sanitation District in Orange County in California with significantly beneficial results. RCM and Enterprise Asset Management best practice for new assets (applied in the design and development phase is a subject in itself. It is noted that experience with initiatives of this nature can result in huge paybacks over the 30-50 year life-cycle of water industry assets.

## COSTS AND BENEFITS OF AN RCM PROGRAM

### Benefits Clearly Demonstrated at Evaluation Sites and in Pilots

The evaluations performed at Broward County and the MWRA show very significant benefits from Reliability projects using RCM which have been implemented in a strategic manner -adhering to the critical success factors identified in Chapter 5 (Recommendations to Utilities on how to implement a Reliability Program using RCM). The MWRA has improved on all its major maintenance KPI's (see Appendix I) while reducing the number of maintenance hours by over 25% and reducing both preventive and corrective maintenance. Broward County recorded a 40% reduction in annual maintenance costs from pre RCM days (average costs 1997-2001) versus post RCM times (average 2002 and 2003).

On the other hand the maintenance program survey has shown some utilities tinkering with RCM and not realizing very significant benefits.

The conclusion drawn from this is that for utilities to realize the potential benefit of using RCM to develop an optimized and effective maintenance program, the RCM initiative or program

- In order to assist the City in determining their capital financial needs and avoid unexpected capital requests during the year, VWI has **consistently identified critical capital projects in the Capital Plan** and has achieved both incentives in this category each year.

TECHNICAL OPERATIONS & MAINTENANCE

VWI has implemented operational and distribution system modifications to improve water pressure to areas of the IW system prone to experiencing low pressure. Using pressure monitors installed within the distribution network, VWI continually monitors pressures at key locations. The number of occurrences in 2003 and 2004 in which the pressure at any of these locations dropped below 30 psi was reduced by an average of over 40% from the number occurring in the baseline year of 2002. This reduction was due to a combination of recommending and implementing distribution system infrastructure improvements in certain critical areas, and ensuring that the pumping system is operated in a manner to minimize even momentary low pressure occurrences.

VWI has made process adjustments in several areas to provide more efficient operation and improved water quality. Modifications were made at the Fall Creek and White River

**Key Highlights**

- VWI has increased efficiency in the system through the use of technology.
- Water quality has been increased with the use of alternative treatments.
- Project management technology has been streamlined for more effective tracking.

to assess potential water quality degradation within the distribution system. A new control room was constructed at the White River Treatment Plant to upgrade the control and reliability of the treatment process. In addition, the control room included an operator interface for controlling pumps and valves throughout the distribution system to provide an alternative location for these operations.

VWI successfully implemented Synergen in September 2003. Synergen is an Enterprise Asset and Work Management system (EAM) / Computerized Maintenance Management System (CMMS) that allows companies to optimize asset performance and reliability. Through the use of the CMMS, maintenance programs are streamlined and the full life cycle of the asset is taken into consideration. Synergen is a well-structured program that allows projects to share information to improve overall efficiency.

Maintenance costs were more accurately captured and reporting to the client was greatly improved. Preventive Maintenance (PM) tasks were generated based off calendar date and/or run time. Initially the PM were developed from past practice, but with the implementation of Reliability Centered Maintenance (RCM), the frequency and scope of the PM are changing for improved asset management.

RCM is a process used to determine what maintenance activities should be done to an asset to ensure it operates at the user's expectations throughout its life. RCM uses seven separate activities to identify what should be done to properly maintain an asset. VWI has analyzed over 20 separate processes throughout the overall system. Many predictive tasks have been established and the results have been favorable.

The process of continually inputting facility information into the water system Global Information System (GIS) has improved since VWI began management of the system. The process has

**Figure 4.1 2004 annual report excerpt – City of Indianapolis, Veolia Water Indianapolis**

must be structured correctly. The two pilot sites and the evaluations bear this out very strongly. Both pilot programs have been successful leading on to strategic programs at the participating utilities. The two established programs evaluated have been strikingly successful and have moved asset management to a different level in the utilities involved.

### Costs to Initiate an RCM Program

The cost of a pilot site is carefully evaluated in Appendixes F and G and in the recommendation specified in Chapter 5. To get started in RCM with the foundations of a strategic program the basic external costs are estimated at $65,000 to $110,000 depending on scale of pilots and number of people to be trained. (Table 4.1 and Table 4.2)

### Benefits To Be Obtained From RCM Programs

The evaluations performed at existing sites, the two pilot programs and research in literature on RCM shows that benefits can be obtained in the areas listed below. These are detailed in Chapter 5.

**Table 4.1**
**Estimated external costs for pilot project, training with internal facilitator**

| Item | Likely unit cost | Quantity required | Estimated cost* | Notes |
|---|---|---|---|---|
| 3 day RCM intro training | $12,000 | 1-2 courses (Training 20-40 staff – up to 20 on each course) | $12,000-$24,000 | More courses required as more personnel trained in large-scale application |
| Consulting mentoring, facilitation, project management | $1,500 | 20-40 days | $30,000-$60,000 | Dependent on intensity and strength of internal team and whether analyses are facilitated by internal resources or not |
| Train internal facilitators | $6,000 | 2 staff trained as facilitators | $12,000 | Unit costs will reduce if greater numbers trained |
| Procure RCM software licenses | $12,000 | 1 five-user license | $12,000 | Based on 5 user license for leading package |
| Estimated total for external costs | | | $64,000 to $110,000 | Depending on scale of pilots and number of people trained |

* Estimate includes travel costs

**Table 4.2**
**Internal resource requirements for pilot project**

| Item | Qty per RCM analysis | # of RCM analyses in pilot | Estimated quantity over 2 month duration | Notes |
|---|---|---|---|---|
| Review group for each analysis | 15 person-days | 3 | 45 person days | 4-5 people for 5-15 meetings for each analysis = 15 man-days per analysis |
| Project Management | | | 30 person days | 0.5 FTE for duration of 2 months |
| Facilitator | | | 40 person days | 1 FTE for Duration of 2 months – normally train 2 |
| Steering Committee | | | 4 hours | 2 Steering Committee meetings of 2 hours |
| Project Sponsor | | | 2 person days | 2 hours per week |
| Audit of results | 1-2 days | | 3 days | Manager responsible for assets |
| Total | | | 115-120 person days | |

FTE = Full-time equivalent

- Greater safety and environmental integrity (improved reliability)
- Improved operating performance (output, product quality and customer service) (improved reliability)
- Greater maintenance cost-effectiveness (as much as 25-30% in established programs evaluated)
- A comprehensive database defining the maintenance program
- Optimized MRO Spares Holdings
- Greater motivation of individuals
- Better teamwork
- Reduced Risk
- Reduced Insurance Costs
- Improved Financing – Bond Rates
- GASB 34 and Asset Management benefits

**Summary – Cost – Benefit Analysis**

The major conclusion of this report is that water utilities will typically derive extensive benefit from utilizing RCM to develop optimized maintenance programs for their asset base. This is well demonstrated in the evaluations of established RCM programs at the two water utilities and in the results achieved with the pilot projects performed as part of this research project. The caveat resulting from the survey and the experience of the authors of this report is that if projects are not properly structured or are approached informally at a low level in the organization, results may vary. This is indicated by the survey where those respondents who use RCM typically showed benefit but to varying extent. Detailed analysis of the established RCM programs in the case studies included in this report (at MWRA and Broward County) both showed substantial cost reductions while simultaneously improving asset performance. At the MWRA substantial cost savings in the order of **25-30%** were achieved by reduction of both Preventive and Corrective Maintenance hours while equipment reliability and plant availability was improved. The case study at Broward county also showed very significant maintenance cost reductions (>25%) from both reduced preventive and corrective maintenance while asset performance was improved.

The conclusion to be derived from the detailed evaluations on case studies of established RCM programs is that benefits from RCM substantially outweigh the costs of implementing the RCM program. The programs provided substantial payback from cost reduction aspects alone – before quantifying benefits of improved reliability and asset performance. *It should be stressed that the objective of an RCM program is to optimize the reliability and performance of the assets concerned by developing the most effective maintenance program. Cost-savings may result from an RCM program dependent on the prior situation but the primary objective of the program should NOT be to cut people or maintenance cost.*

## KEY RCM SURVEY RESULTS

A total of 177 respondents representing water utilities across the United States and in Canada completed an online survey that was developed by New Dimension Solutions. The survey instrument was designed to establish personal data of the respondent, develop a profile of the utility that the respondent represents, determine maintenance programs and efforts currently made within their facilities, and ascertain their involvement with and any benefits achieved through the application of Reliability-Centered Maintenance (RCM).

The survey participants included 177 respondents with a representative mix of utility sizes for <10 MGD to >300 MGD. The majority of the respondents were city- or municipality-owned (77%). The majority performed their own maintenance (>85%).

## Current Maintenance / Asset Management Performance and Priority

Data revealed that there is an emergent focus by utility management on asset performance and optimal management of the utility's assets. Over half of respondents (57%) indicated that this focus is becoming an important issue for utility management, and 13% of respondents stated that optimal management and performance of the utility's assets is becoming ***a very critical issue for utility management***. Respondents were also asked to provide reasons for any change in management prioritization of physical asset maintenance and performance. When the first two reasons for this change in management prioritization were grouped together, aging equipment, facilities, or infrastructure (21%), the desire for better performance, higher efficiency, or optimization (16%), the need for cost reduction or to control costs (10%), and for compliance with regulatory requirements or more stringent regulations (10%) were the most commonly mentioned responses.

Respondents were also asked to specify the most serious issues that currently face utility management, in terms of the maintenance of its assets. When the top three most serious issues were grouped together, capital replacement costs (45% included this in their top three) was the most commonly mentioned, followed by maintenance costs (44%), aging facilities (42%), and reliability (40%).

To gain an understanding of the manner in which maintenance departments function across the various utilities, respondents were asked to select the mode of operation of their maintenance department. A majority of respondents (61%) indicated that their maintenance department function operates in a steadily improving proactive mode, recognizing that some asset systems and facilities continue to be a problem. Twenty-two percent (22%) of respondents say that their maintenance department function operates with the basics of a proactive program, but that maintenance is still mostly reactive in nature.

Plant breakdowns are defined in the survey as being unscheduled failures of an asset that affect plant operations and must be rectified immediately when the asset fails. This is an extremely important issue for utilities with highly loaded or critical assets. According to the survey, on average, 20% of maintenance work is in direct response to these plant breakdowns.

Managing assets via a Computer Maintenance Management System (CMMS) or an Enterprise Asset Management (EAM) system involves tagging physical assets and registration of these assets within the system. The degree to which utilities tagged physical assets and registered these assets within the CMMS / EAM system differed across respondents. For example, 35% of respondents said that the tagging and registering of some critical assets have been completed, yet there was still much more work to be done. Thirty-one percent (31%) indicated that the tagging and registering of most assets has been completed, including all critical assets, with some clean-up work still to be done. ***Another 31% said that no assets at all have been tagged or registered.***

Fixed-interval maintenance tasks (also known as preventive maintenance tasks) are developed from several methods including OEM recommendations, equipment history, formalized preventive maintenance optimization processes, and RCM. When asked to specify the percentage that each of these methods have contributed to the development of the fixed-interval tasks at their utility, respondents on average assigned 48% to OEM recommendations, 31% to equipment history, 10% to formalized preventive maintenance optimization process, and 3% to RCM.

## Asset Management Programs – Key Points From Survey

Asset management is becoming an important issue with 70% of water utilities. The majority of water utilities believe there is room for improvement in their current utility asset management program and that they are steadily working to improve their programs. Senior management commitment to asset management was developing or strong for greater than two thirds of respondents.

The asset management program for the majority of utilities is in the starting or initial phases with another 22% with no asset management program. The majority of critical assets were included in the asset management programs. The utilities considered that the overall asset condition was the same or improving at over 75% of all utilities with 22% stating that asset condition was deteriorating.

The top three reasons for the formation of an asset management program are:

- Aging infrastructure
- Better performance
- Cost-savings/Reductions

## Maintenance Programs / Failure Management Policies

Formal and systematic processes and methods to derive maintenance programs generally appear to be in place in many facilities. Forty-five percent (45%) of respondents indicated that these methods have been deployed or utilized at their utility on certain critical assets, and 14% said that a comprehensive program utilizing formal processes to derive asset maintenance programs is in place across the utility. Another quarter (26%) stated that there is a sporadic effort to implement formal and systematic processes / methods at their utility, and 15% said that there are no formal or systematic processes / methods in place at the utility to facilitate the development of maintenance programs for the utility's assets.

Respondents were asked to indicate their level of familiarity with the main Maintenance Strategy/Program Formulation Techniques currently available for developing optimized maintenance programs for physical assets. Familiarity with Original Equipment Manufacturer (OEM) Programs was highest (74%) followed by PM Optimization (46%), RCM (36%), and Root Cause Analysis (34%). In addition to these, respondents were asked to specify which maintenance strategy formulation techniques were used to develop maintenance programs for critical and non-critical physical assets. In both of these cases, the Original Equipment Manufacturer Program was used on the highest percentage of physical assets, followed by PM Optimization, and then RCM.

Applying formalized maintenance strategy formulation techniques or reliability optimization methodologies during the design-build phase of new assets can optimize future maintenance of assets and minimize life-cycle costs. Respondents were asked whether such Maintenance Strategy/program Formulation or Reliability Optimization techniques are applied at their utilities either during the design-build phase of new assets, or during the modification of existing assets. Approximately 44% of respondents indicated that the maintenance staff performs some formalized reliability review of the proposed design, and 31% indicated that design engineers and draughtsman try to consider reliability and maintainability during design. About 22% stated that no formalized maintenance strategy formulation techniques or reliability optimization methodologies are applied in the design-build phase or when modifying existing assets.

With respect to formalized Criticality Analyses for the technical systems or assets within the utility facilities, 44% of respondents indicated that no such analyses have been performed at their utility. One third of respondents (32%) stated that this type of analysis has been performed for some areas of the plant, and one fifth (20%) said that a formalized Criticality Analysis has been performed for most areas of most facilities.

An effective reliability program includes activities such as regular reviews and adjustments to the maintenance program and standard operating procedures. Nearly half (47%) of applicable respondents indicated that some asset failures and performance shortfalls are reviewed to identify cause and to then develop appropriate failure management policies, and 29% indicated that asset failures and performance shortfalls are reviewed for critical equipment.

Reviews of asset maintenance programs appear to be a common task among respondents, as over half (52%) indicated that some of these programs are reviewed on a regular basis. About 22% indicated that these programs are reviewed on a regular basis for critical equipment, and 12% said that all of these programs are reviewed on a regular basis. When maintenance programs are developed, over half (52%) of applicable respondents stated that some of the process steps used to develop the maintenance programs are formally documented. Nearly one quarter (24%) said that none of these steps are documented.

Earlier, it was noted that reliability is among the top three most serious asset maintenance issues facing utility management. Given the severity of this issue it is no surprise that there is an emergent awareness among senior management with respect to their commitment to reliability initiatives. More specifically, 39% of respondents feel they have strong senior management commitment towards reliability improvement initiatives, and 29% feel management is aware of these initiatives. When asked whether or not their utility has any formalized strategic programs in place to improve reliability of the utility's physical assets, 30% of respondents indicated that their utility is starting a formal program to improve asset reliability, and one quarter (25%) said that there are initial discussions and plans to commence such a program.

Approximately three quarters (74%) of respondents said that their utilities do not have a reliability group in place at their utility. One fifth (21%) of respondents said that an engineer on a part-time basis fulfills the reliability group function.

Utilities generally appear to be making an attempt at keeping up-to-date with new initiatives and industry standards. More specifically, 37% of respondents indicated that their utility takes part in some benchmarking and networking activities and 27% said that their utility belongs to organizations.

## Reliability-Centered Maintenance (RCM)

Of the 177 valid utility responses, 42 went through the section on RCM. [Participation in this section was limited to utilities "where RCM has been used in a significant manner"].

*Key Points on RCM From Utility Survey*

The majority of water utilities use OEM recommendations to formulate maintenance strategy for critical and non-critical assets. RCM is not yet widely used by the water utilities industry (only 10% of the 177 respondents responded to the majority of the RCM questions.)

83% of 18 utilities recognized a positive payback on the RCM effort, with five utilities stating the payback was within 3 months to two years.

The major benefits from the RCM program were stated to be:

- Knowledge Retention
- Asset and System Knowledge
- Improved maintenance work completion
- Reduced maintenance cost
- Operations and Maintenance Co-operation

*Other Salient Data Points on RCM in Utility Survey*

Respondents from those utilities that use RCM were asked to indicate the extent to which RCM has benefited the utility. Knowledge retention for the utility was the most commonly mentioned benefit (60%). Knowledge of assets / systems was another commonly mentioned benefit (58%) along with improved completion rate on maintenance tasks, and reduced maintenance cost (54%). Regarding reliability, 16% saw significant improvement and 32% saw improvement in some cases.

Benefits were surveyed for thirteen variables. The sample of respondents on benefits ranged from 23 to 25. Across the thirteen variables surveyed, the majority of respondents saw benefit raging from marginal to significant, with approximately one-third seeing no benefit.

In terms of their importance to the utility, respondents ranked reliability as the most important benefit, followed by cost reduction and then knowledge retention.

Derivation and calculation of required frequencies for failure finding tasks for hidden failures is performed in some RCM methodologies (32%). Fully 40% weren't sure whether or not the RCM Methodology used includes this type of calculation.

Respondents were also asked to state their level of awareness with SAE Standard JA1011, which defines standards for acceptable RCM processes. Among applicable respondents, only 12% are familiar with it, and 14% have heard of it, but are not familiar with the details.

## Potential Benefits From Use of RCM Based on Survey Results

*Duty-Standby Strategies*

Utilities capacity was less than demand for 97% of the utilities surveyed (i.e. the utilities had some excess capacity). The RCM process uses duty-standby strategies to extend equipment life and improve asset reliability. The excess capacity of utilities in normal load conditions, can be leveraged to provide maintenance cost savings and improve equipment reliability by using optimized duty-standby strategies.

### *Reduction of Emergency/Breakdown Work*

The failure of assets that affect plant operations that must be rectified immediately (emergency work) exceeds 10% of all work at the majority of utilities. The industry benchmark for emergency work is <5%. The use of RCM can significantly improve equipment availability and assist in improving plant performance by the use of standby-duty operation, improving the preventive maintenance of asset components by the use of condition monitoring techniques and the elimination of counter-productive, intrusive tasks and identifying appropriate maintenance tasks/functional checks for hidden failures.

### *OEM-Recommended Maintenance Programs Implies Opportunity for Optimization*

The preventive maintenance programs for approximately half of the utilities surveyed are based on OEM recommendations. The use of OEM-based maintenance programs does not account for the operating context of the equipment in use. The operating context of each plant's equipment (assets) is reviewed during the RCM analysis process and considers the actual operating conditions such as flow, pressure, temperature and process fluid used by each component and its effect on the optimal maintenance required by that asset in those conditions. For example, a sump pump that operates infrequently and is backed up by standby unit may be run to failure, while a critical raw waste water pump that runs 24/7 may require extensive condition monitoring to optimize its availability and avoid consequences of unexpected failure. Plainly the OEM recommendations cannot fit every operating context – even if they have been compiled based on Reliability best practice and RCM concepts (which is frequently not the case). One size does not fit all. Thus this finding that over 50% of maintenance programs are based on OEM recommendations implies that there is much scope for optimization of maintenance programs using RCM.

### *Lean Optimized PM Program Leading to High Completion Rates*

The survey results indicate that the preventive maintenance tasks at water utilities are scheduled approximately half the time (48%) and 59% of these tasks are completed. The correct scheduling and completion of preventive maintenance tasks is important to ensure equipment reliability and availability. The establishment and completion of "PMs" is one of the basic building blocks of an asset management program. The industry benchmark for completion of PM tasks is 90%. RCM can improve the preventive maintenance program performance by eliminating low-value PMs, correcting the frequencies on PMs which are set or based on incorrect premise and ensuring that only required PMs based on sound concepts remain. Implementing RCM based on the recommendations laid out in this survey also ensures that maintenance and operations personnel associated with the asset take ownership of the PM program. Thus RCM programs lead to high completion rates by leaning the PM program to an optimized set of effective tasks while simultaneously building ownership of the program with the people that do the PM tasks and operate the assets.

## Other Asset Management Issues

### *Asset Tagging*

The first step in any asset management program is to identify all assets to be managed/maintained at the utility. Asset tagging using a unique identifier is required to do this to link information to and track activity against that asset. Two thirds of respondents stated that some critical assets were tagged and that much work needed to be done to complete tagging.

### *CMMS System*

Implementation of a computerized maintenance management system (CMMS) is often the next step in the development of an asset management program. CMMS are often not used for small utilities due to the lesser number of assets and size of maintenance staff. The majority of utilities (54%) use a CMMS. Those using CMMS include > 70% of assets in the CMMS database. The average use of the CMMS was five years.

### *Asset Criticality Rating*

The criticality of assets was not established for approximately half of all utilities (44%). The establishment of asset criticality is an important step in work prioritization, budgeting and implementation of an asset management program.

# CHAPTER 5
# RECOMMENDATIONS TO UTILITIES

## COMPONENTS OF AN SAE-COMPLIANT RCM PROGRAM, AND HOW THEY COULD BE APPLIED IN THE WATER INDUSTRY

The SAE standard JA1011 defines the '*minimum criteria that any process must comply with to be called "RCM." It does not attempt to define a specific RCM process.*'

The SAE notes the following about the SAE JA1011 standard. '*It is intended for anyone who wishes to ascertain whether any process that purports to be RCM is in fact RCM. It is especially useful to people who wish to purchase RCM services (training, analysis, facilitation, consulting or any combination thereof.)*'

Section 5 of the SAE standard specifies that "*Any RCM process shall ensure that all of the following seven questions are answered satisfactorily and are answered in the sequence shown as follows*:

1. What are the functions and associated desired standards of performance of the asset in its present operating context (functions)?
2. In what ways can it fail to fulfill its functions (functional failures)?
3. What causes each functional failure (failure modes)?
4. What happens when each failure occurs (failure effects)?
5. In what way does each failure matter (failure consequences)?
6. What should be done to predict or prevent each failure (proactive tasks and intervals)?
7. What should be done if a suitable proactive task cannot be found (default actions)?

The standard specifies that to answer each of these sections "satisfactorily" that certain specific information should be obtained. This is defined in Sections 5.1 through 5.8 of the SAE Standard. These can be referenced in the SAE standard JA1011 which can be obtained from the SAE (www.sae.org). Copies of the SAE Standard JA1011 may be ordered from the SAE through the SAE web-site (www.sae.org) or at the following contact numbers (724) 776-4970 or Fax (724) 776-0790.

The SAE standard also specifies (in Section 5.9) that **periodic review** is required of an RCM-derived asset management program to ensure that the program is sustained as optimal and valid. The requirement for a "living RCM program" recognizes that (a) much of that data and estimates used in the original RCM analyses are imprecise and may be improved or refined over time (b) that assets and operations are modified, performance expectations and other contextual conditions, change and (c) that maintenance and asset technology improves/evolves over time.

The RCM process used in the two pilot programs and in the two detailed case studies is RCM2 – developed by Aladon Inc. under the direction of John Moubray. The RCM2 product fully complies with the SAE standard JA1011. The reports in the case studies and on the pilot programs indicate in detail how these SAE-specified components can be applied at water utilities.

## RECOMMENDATIONS ON UTILITY APPROACH TO IMPLEMENT AN RCM PROGRAM

The recommendations laid out below have been developed from experience of the authors, analysis of successful case studies and recommendations from leading providers of RCM process-ware. They can be considered as critical success factors in the development of a successful, sustainable reliability program based on adoption of RCM as an underlying philosophy and maintenance strategy/program formulation process.

### Senior Management Ownership / Sponsorship for the RCM Program

Real success using RCM will only be achieved as a strategic program driven from the top of the organization. Isolated groups may achieve positive results in analyzing specific asset systems, but structural, systemic change will only be achieved with senior management ownership of and commitment to the program. The reasons for this include the following:

Sustainable reliability programs require concerted change management and training programs which require significant commitments of funds and key people's time. Many of the tenets of RCM require a fundamental shift in the way people think about maintenance. Such systemic change will typically only occur if there is this commitment from the top of the organization.

### Structure the Reliability Initiative as a Strategic Program With Defined, Measurable Business Objectives

Adoption of RCM or development of a reliability program is of course not an end in itself. Such initiatives are adopted and implemented to deliver a positive return or payback to the organization. In many cases asset-intensive organizations are driven to implementing Reliability programs by specific problems or competitiveness issues in their business operations. In other cases management may have read of or heard of RCM success stories at conferences or similar events and perceived the opportunity for improved effectiveness in their own asset management set-up.

It is highly recommended that Reliability programs based on RCM are structured with specific formalized and measurable business objectives. This should be the case for both pilot projects and for the roll-out of the full program. This approach is critical for a number of reasons. Structuring the program to achieve business objectives that are in line with the organization's strategic goals ensures relevance and attention of senior management. Demonstrating success from pilot projects against the metrics of the defined business objectives will ensure funding of the core project.

In asset-intensive organizations one is typically confronted with large numbers of assets and opportunities for improvement. Effort must be focused where it will achieve the most significant return per unit of resources (the "biggest bang for the buck"). If RCM projects are set up with carefully defined, measurable objectives which are strategically congruent with the organization's business goals, then effort will be focused where it matters i.e. on those assets where the need is greatest in terms of meeting business objectives.

## Select a Reputable Experienced Consulting Partner to Assist in Initiating the Reliability Program

Successful reliability programs have moved from an art to a science. Following a proven RCM approach will definitively yield results (as evidenced by the evaluations of existing programs at the two participating water utilities and the two documented case studies for the pilot projects).

As is noted below there are a number of activitics that must be well planned and executed to ensure a successful reliability program based on RCM. These include well-designed effective change management and training plans and an effective approach to actually doing RCM (as opposed to understanding the concepts). It is a consistent experience that programs which are initiated with the help of experienced consulting firms, who specialize in RCM, are far more likely to succeed than projects where organizations try to tackle an initiative on their own based on reading the literature. Utilizing experienced consulting assistance can help avoid falling into known potholes and give the program critical impetus in its infancy.

Specialized firms can assist with items including the following:

- Project structuring and approach
- Selection and planning of pilot projects
- Determination of asset criticality
- Project Management
- Training and Change Management (review groups, facilitators, project management, reliability engineers, management, other stakeholders
- Selection and profiling of key project personnel (project management, reliability engineers, facilitators etc)
- Mentoring of organizations own facilitators and structured improvement of facilitator competency
- Contract facilitation
- Robust practical RCM processes
- Software to support the RCM analysis
- Review of RCM analyses
- Implementation of RCM results
- Review of Maintenance and Reliability organization structure
- Design and implementation of revised business processes reflecting reliability focus
- Review of integration between RCM program and execution system (CMMS/EAM system)

Plainly there are many areas involved in correctly and optimally structuring a Reliability program to achieve business results. It is sound business sense to utilize expert advice on how to do this. The philosophy of utilizing such expert help should be on maximizing knowledge and skill transfer in-house so that the utility becomes proficient and largely self-sufficient in the core skills and processes of a successful program and that use of specialized help is reduced over time to mentoring, review and audit of critical or problem aspects of the program.

### Base the Program on a Reputable RCM Methodology Which Is Compliant With the SAE Standard JA1011

As a search on the internet will indicate, there are currently a plethora of "RCM" service and software vendors in the market-place, many promising the "next best thing" in terms of adaptations of the RCM process. This can be confusing and daunting to a company considering an RCM initiative.

The recommendation here is to utilize services and software from vendors who are fully compliant with the SAE standard. The SAE JA1011 standard is not a long complicated document. It is a brief, clear and succinct statement (less than ten pages) of the criteria that any process must comply with to be called RCM. As noted in the section at the beginning of this chapter, the heart of the process is the seven questions or steps of RCM. These are at the heart of all reputable RCM processes.

The variations on the core RCM process amongst the reputable service and software vendors lie in different approaches and emphasis in how RCM is implemented – in areas such as who does the analysis; change management and training approaches; depth/quality of training materials; treatment of areas such as risk management, the development of failure finding tasks for hidden failures and the statistical models supporting these areas; software systems to support the process.

Adherence to the SAE JA1011 standard rules out the various vendors who are offering "streamlined RCM" approaches, Planned Maintenance Optimization (PMO) variants and any other process which does not include the seven questions specified in the standard. These processes cannot be called RCM and they dangerously compromise the RCM process. If a utility is going to do RCM it should absolutely use a quality process which is compliant with the SAE standard. Engineering is a discipline where ill-informed compromise can lead to very serious consequences. Nobody in their right minds will compromise on safety in asset design, for example. By the same logic no compromise is acceptable in the process used to develop maintenance programs on such assets. The criteria specified in the SAE standard define a process which is scientifically sound for managing consequence of failure. Short-cutting or omitting any of these steps is not acceptable.

Plainly one cannot perform RCM analyses on a complete, existing facility overnight. An RCM initiative must be a structured systematic approach prioritized by asset criticality. It is noted that some of these other methodologies may have a place in minimizing risk or improving the status quo in lower criticality areas of the plant before the RCM program rolls out to cover those assets – they should not, however be considered as a substitute for RCM for any asset systems where the consequence of failure is significant in terms of safety, the environment, operational or other impact.

### Include Effective Change Management and Training Into the Approach

To ensure that the organization implements a reliability program which is successful (in achieving its defined objectives) and self-sustaining (continues to build and improve over time) it is vital that the people impacted by the program understand why they are doing the program activities and that they believe in the process. They must be committed to the program and own the results. These are **change management** issues. For many people involved in maintenance and operations of asset-intensive organizations (at all levels of the organization), embracing the concepts of RCM typically involves some fundamental rethinking and deep-seated changes to their understanding of aspects such as how assets fail and how they should be maintained. Unless these stakeholders go through effective hands-on training on the RCM process which fully explains and illustrates the concepts, they are likely to resist the program and create serious resistance-to-change issues. **Understanding and involvement breeds commitment.**

This implies that the change management process must include appropriate training on the RCM process for everybody who is significantly involved in the program (maintenance, operations, engineering, reliability and specialist personnel – possibly OEM representatives or engineers from consulting engineering firms). This training breeds a common understanding and language among these different functions in the organization and frequently works wonders for co-operation on many other issues beyond the RCM program.

It is however not enough to train all the stakeholders in the asset management organization. It is our strong belief that the stakeholders relevant to each asset system are involved in the RCM review process. There are several benefits to this approach. From a change management perspective involvement of people in the process means they are informed and involved. As noted in the section below (Review group approach) it is our recommendation that the review group performs the analysis (under the guidance of the facilitator in terms of process) and makes the decisions. This means that they own the results – the RCM recommendations are not a set of foreign tasks parachuted into the organization from external consultants. This is critical from a change management perspective. The people actually doing the maintenance work (who know the assets and how they fail), the operators of the asset system and appropriate technical specialists are the people who actually do the analysis. This is covered in more detail below.

In summary from a change management and training perspective organizations must make the investment to train all appropriate stakeholders on the RCM concepts and process to a level that they really understand the theory, language and hands-on application of RCM and that they "buy into" the process. In addition the RCM process on site must utilize a review group concept which involves the stakeholders for each asset system reviewed (including maintenance, operations and appropriate specialists. These stakeholders should make the decisions on a consensus basis (guided through the process by a trained facilitator) so that they own the results.

It is our strong recommendation that ramming in recommendations from RCM analyses performed by external engineers, consultants is highly problematic and frequently results in failure or sub-optimal results.

Plainly, where RCM is performed up front in a design environment (before the maintenance and operations organizations are present, the make-up of such review groups may be somewhat different. The principles developed above still hold good. Stakeholders in the process should be trained in RCM to a deep level of understanding and involvement of appropriate personnel must be maximized (design and process engineers, technical specialists and ideally key people who will be part of the maintenance and operations groups in the operating phase of the asset life-cycle.)

### Use "RCM Review Group" Approach

An "RCM review group" approach is highly recommended for both reasons of "ownership" of the RCM analyses and to ensure optimal access to knowledge available to perform the analysis.

The term review group is used to describe the group of people who actually perform the RCM analysis. The recommendation for existing facilities is that such review groups consist of:

- Trained RCM Facilitator
- Maintenance representative(s) (from the maintenance group that actually maintains this asset system)
- Operations representative(s) (from the operations group that actually operates this asset system)
- Specialists as required (possibly from OEM, process engineering, specialized technical areas, condition monitoring expertise, etc.)

For new facilities that are still in a design stage the composition will change to reflect which personnel are available.

It may appear more time-consuming to follow a review group approach but it is far more effective and takes less time in the long run. We have seen various organizations try to perform RCM analyses using third party contract engineers or internal Reliability service departments who analyze drawings, P and ID's etc in isolation. This approach is ineffectual and incomplete at best and can be very dangerous. Incomplete process and design understanding is achieved doing RCM analyses just by studying asset documentation in isolation – as can readily be demonstrated even on relatively simple systems. This third-party "contract" RCM approach typically has low levels of acceptance with the internal maintenance personnel and results in the PM program having low credibility and completion rates.

The recommended review group approach involves local maintenance personnel who have experience of the asset system under real operating conditions and are familiar with its operation and failure modes. Involvement of these personnel changes the exercise from theory to practicality.

The review group process also has other very fundamental benefits. Experienced RCM practitioners consistently report that it is only during review group sessions that maintainers and operators often get to really understand how systems work. This is especially true for control systems and protective devices. It is also very frequently the first time that maintenance and operations personnel really communicate properly and with a common language.

### Select Internal RCM Facilitators Carefully According to Correct Criteria

It is important to select the internal RCM facilitators carefully to ensure credibility and sustainability of the RCM program. Reputable RCM consultants will typically have a tool to support this facilitator selection process. Criteria include:

- Ability to communicate
- Ability to think logically and systematically
- Technical background and experience
- Project management and organization capabilities
- Ability to be a team player
- Cultural fit and acceptance in the organization
- Availability
- Responsiveness to new ideas
- Any "baggage" from other initiatives
- Computer literacy

## Initiate the Program With a Pilot Project Including Several Analyses

It is recommended to begin the RCM program with a proof-of-concept consisting of one or several pilot projects with RCM analyses on targeted asset systems (Table 5.1). This enables senior management to assess the results that can be obtained with RCM without committing major resources. A typical time-line for a proof-of-concept program is shown in Figure 5.1 with a Gantt chart view.

**Table 5.1**
**Possible sequence of activity for the RCM proof-of-concept (pilot)**

| Activity | Note |
|---|---|
| Project Start | |
| 2-3 Hour Presentation for key Stakeholders, Project Mgr" | Ensures stakeholders understand what the project is about |
| Plan Pilot Project(s) | Select critical or problem system(s) |
| Specify Pilot Site Business Objectives | |
| RCM 3-Day Intro Training for Key Stakeholders, Project Mgr, Review Group Members | More in-depth training for sponsor, management in sections where assets are located, review group etc |
| Select Facilitators | Pilot project can be facilitated by consultant |
| RCM Facilitator Training | |
| Prepare for RCM Pilot projects | |
| RCM 3-Day Intro Training for any untrained Review Group Members (if required) | |
| Procure RCM Software or arrange trial copy | |
| Perform RCM Pilot Analysis 1 | |
| Perform RCM Pilot Analysis 2 | |
| Perform RCM Pilot Analysis 3 | |
| RCM Consultant to Mentor Facilitators | |
| Prepare Management Presentation and High-Level Roll-out Plan | |
| Compare results versus short-term objectives | Compare results against objectives – longer term objectives can only be reviewed later |
| Deliver Management Presentation | |
| RCM2 Pilot Projects Completed - Milestone | |

| ID | | Task Name | Duration | Start | Finish | Predecessors |
|---|---|---|---|---|---|---|
| 1 | | Concept RCM High-Level Project Plan - Proof-of-Concept Ver 1 | 76 days | Fri 9/10/04 | Fri 12/24/04 | |
| 2 | | Project Start | 1 day | Fri 9/10/04 | Fri 9/10/04 | |
| 3 | | 2-3 Hour Presentation for key Stakeholders, Project Mgr | 1 day | Mon 9/13/04 | Mon 9/13/04 | 2 |
| 4 | | Plan Pilot Project(s) | 5 days | Mon 9/13/04 | Fri 9/17/04 | |
| 5 | | Specify Pilot Site Business Objectives | 2 days | Mon 9/20/04 | Tue 9/21/04 | 4 |
| 6 | | RCM 3-Day Intro Training for Key Stakeholders, Project Mgr, Review Group | 3 days | Wed 9/22/04 | Fri 9/24/04 | 5 |
| 7 | | Select Facilitators | 2 days | Mon 9/27/04 | Tue 9/28/04 | 6 |
| 8 | | RCM Facilitator Training | 10 days | Mon 10/4/04 | Fri 10/15/04 | 7 |
| 9 | | Prepare for RCM Pilot projects | 5 days | Mon 10/18/04 | Fri 10/22/04 | 8 |
| 10 | | RCM 3-Day Intro Training for any untrained Review Group Members (if requ | 3 days | Mon 10/25/04 | Wed 10/27/04 | |
| 11 | | Procure RCM Software or arrange trial copy | 1 day | Thu 10/21/04 | Thu 10/21/04 | 8 |
| 12 | | Perform RCM Pilot Analysis 1 | 20 days | Mon 11/1/04 | Fri 11/26/04 | 11 |
| 13 | | Perform RCM Pilot Analysis 2 | 20 days | Mon 11/8/04 | Fri 12/3/04 | 9 |
| 14 | | Perform RCM Pilot Analysis 3 | 20 days | Mon 11/15/04 | Fri 12/10/04 | 9 |
| 15 | | RCM Consultant to Mentor Facilitators | 30 days | Mon 11/1/04 | Fri 12/10/04 | 12SS,14FF |
| 16 | | Prepare Management Presentation and High-Level Roll-out Plan | 5 days | Mon 12/13/04 | Fri 12/17/04 | 15 |
| 17 | | Compare results versus short-term objectives | 3 days | Mon 12/20/04 | Wed 12/22/04 | 16 |
| 18 | | Deliver Management Presentation | 1 day | Thu 12/23/04 | Thu 12/23/04 | 17 |
| 19 | | RCM2 Pilot Projects Completed - Milestone | 1 day | Fri 12/24/04 | Fri 12/24/04 | 12,13,18 |

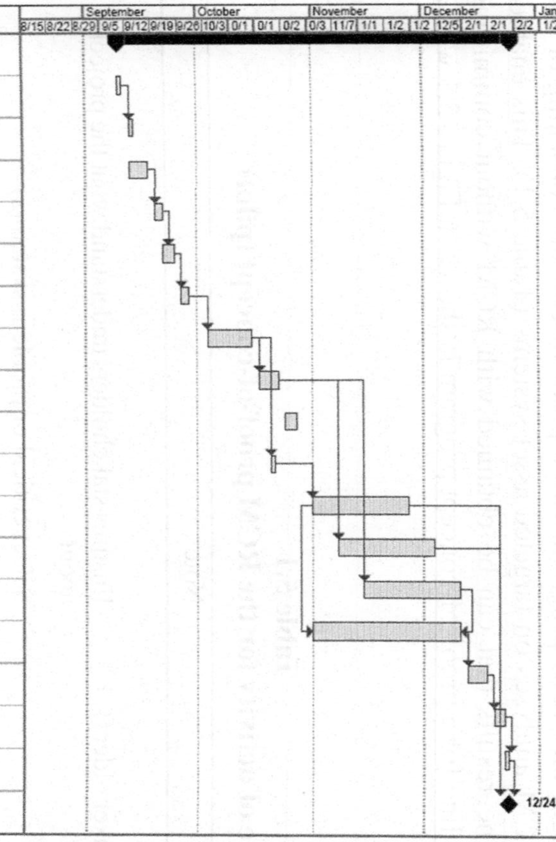


Project: RCM Proof-of-concept Project
Date: Wed 8/31/05

Task · Split · Progress · Milestone · Summary · Project Summary · External Tasks · External Milestone · Deadline

Page 1

Figure 5.1 Outline Gantt chart for proof-of-concept or pilot RCM program

## Capture Analysis Results With Proven RCM Software

The RCM project should be supported with proven software to capture and document the analysis results as each analysis is performed and maintain the audit trail for future review. There are several products on the market from RCM process-ware vendors that are compliant with SAE JA1011.

The RCM process is data intensive and good software which really supports the RCM process is critical to support each analysis and enable "templating," version control and publication of analysis results. The end result of a well-performed RCM analysis is a fully optimized maintenance program - for the specific asset system in its current operating context. It is an auditable, defensible record of why the asset is being maintained in a specific manner. Plainly these records are very important and must be carefully managed on the utility's computer networks.

The RCM software should be a strategic component of the utility's IT landscape for Enterprise Asset Management (EAM). As noted below, the IT landscape for the organization (and the RCM software) should support publication and ongoing access to the appropriate members of the organization.

## Use "Templating" Appropriately

Initial or base-line analyses should be performed for each class of critical asset system. These may be expected to take 5–10 review group meetings, dependent on complexity. These can then either be replicated where subsequent assets and operating context are exactly the same or "templated" where asset and operating context vary. Templated analyses take on average between 10–20% of the time of a full base-line analysis depending on extent of variation.

Water treatment plants and pumping stations are likely to be good environments for templating as there are often numbers of similar asset systems. Thus once a good population of base-line analyses is completed, productivity can be higher on subsequent units. However one must not lose sight of the fact that these units operate within a connected system with other circuitry, control gear, instrumentation, PROTECTIVE DEVICES, metering etc. These aspects of the asset system must not be neglected in the RCM analyses (PART OF THE ASSET"S OPERATING CONTEXT).

Please refer to the case study from the MWRA for additional information on "templating" in a real world situation.

## Make RCM Analysis Results Widely Accessible on a Utility Intranet

RCM analyses are not meant to be done once and then left to molder on a shelf in a file. They should be made readily accessible to the maintenance and operations personnel that work with those asset systems. As noted below an RCM program is a living program where analyses should be reviewed on a fixed –interval basis or whenever relevant change occurs.

The recommendation is that the analysis results are published in a read-only format over the utility's intranet (portal accessible only to utility's own employees) for ready access by maintenance, operations, engineering and reliability personnel.

RCM analyses are an excellent tool to capture knowledge of experienced members of the workforce and publication of the RCM analyses enables transfer of this knowledge to new or less experienced employees.

### Establish an Effective Measurable Process for Implementing Results

A reliability program based on RCM is a strategic program which, over time, requires significant investment of resources. As with any such initiative, payback should be carefully measured. This means gathering required information on the pre-RCM situation to the extent that it exists. Organizations often omit to perform a detailed analysis of the pre-RCM situation to enable an effective comparison.

### Build a Living RCM Program

Change is a constant in every human endeavor. This is particularly true of any significant asset system or facility. It is therefore critical that the RCM analyses and maintenance programs are continually updated. As noted in relation to the SAE standard, the requirement for a "living RCM program" recognizes that (a) much of that data and estimates used in the original RCM analyses are imprecise and may be improved or refined over time (b) that assets and operations are modified, performance expectations and other contextual conditions, change and (c) that maintenance and asset technology improves/evolves over time.

It is recommended that review of the RCM analyses on all assets of a certain criticality level are scheduled as fixed-interval tasks in the EAM system and that RCM reviews are part of the standard procedure for any change/modification or new design.

### Utilize RCM Early in the Life-Cycle of New Assets or Modifications to Existing Asset Systems

Asset-intensive organizations often think of RCM in relation to their existing asset systems and RCM certainly does drive immense value around existing assets. It should be noted however that the most powerful application of RCM is in the design phase of new systems or modifications to existing asset systems. In the design phase the process and asset system designs can be optimized for reliability and protective systems can be designed with maintainability in mind. The revolutionary application of RCM in the air transport (civil aviation) was applied in the design and development phase of an aircraft with truly spectacular results (close to 100× improvement in equipment reliability).

Application of RCM/Reliability and EAM Best practice in the design phase is a subject in itself. ***It is strong recommendation that RCM is a fundamental part of the design and development of new assets and in the modification of any current assets***. It gives structure and a robust, scientific capability to the design process to ensure that designs are optimized for reliable operation. This is very powerful and pragmatic. Payback on investment in RCM and EAM activity in the design phase can be astounding when considered over the life-cycle of the plant.

## BUSINESS RULES FACILITATING IDENTIFICATION OF CRITICAL ASSETS

It is a well-established model in analyzing risk to assess the following:

1. The **consequence** associated with an event (by establishing some form of comparative yard-stick for measuring such consequence which enables the calibration of disparate consequences such as financial, safety, and environmental impacts)
2. The **relative likelihood** or consequence of such events occurring.

The **relative risk** is then assessed by multiplying the two numbers together to establish the relative risk of each event. Attention is then focused on the highest risk events. These can be further prioritized by assessing the relative level of effort required to manage risk for each event and tackling those high risk items which require the least effort first i.e. those with the highest return in terms of risk reduction for effort.

So these concepts can be applied to establishing a critical asset ranking for utility assets. When a utility is commencing a Reliability Program one of the early steps should be to perform an Asset Criticality Analysis. This may be done as part of the Proof-of-Concept phase or as part of the large-scale application of RCM, depending on the scale of the Proof-of-Concept stage (Stage 1).

## Reliability Program Step 1 – Perform Criticality Analysis

As noted above, it is logical to focus initial RCM efforts where they are most needed. This task entails first of all ranking the organization Assets (Equipment) by "relative risk" they pose in preventing the attainment of business goals and secondly, prioritizing the order in which critical assets should be addressed by RCM.

Asset reliability improvement projects compete with the many other improvement programs requiring funding and people. If a utility is to be successful in deploying RCM, it is essential that a business case is developed to justify the expenditure of resources. Management must believe that the potential benefits to the organization's bottom line performance and risk profile justify the investment. The challenge is to rise to the opportunity this represents. As a result, it is recommended to use an Asset Criticality Analysis evaluation tool for this purpose.

The *Asset Criticality Analysis Evaluation Tool* has two purposes. Firstly, to rank Assets by the "relative risk" they pose in preventing the attainment of business goals and secondly, to prioritize the order in which critical assets should be addressed by RCM. Achieving this requires consensus between stakeholders on the result. More precisely, the right information has to be collected from the right people. The Asset Criticality Review group should include operations, maintenance and technology members; at the management, supervision and first line level. Collectively they process the knowledge and responsibility to deal with the question:

> "Where do we need to focus our Asset reliability improvement efforts to maximize contribution to business goals?"

Having determined the "relative risk" rating of the organizations asset systems, the next step looks at the "level of effort" required to improve the reliability of the critical asset systems identified above and to ensure that the maintenance program manages risk in accordance with the organization's profile of acceptable risk. In this case "level of effort" reflects the number of RCM review group meetings required to analyze each asset system. The level of effort numbers are plotted against the relative risk numbers to define the areas of initial focus, which should be directed towards items of high risk, low effort or low cost. Projects with the potential to deliver the maximum benefit to the company by mitigating risk are selected to be executed first.

It is noted that the Asset Criticality Analysis is a focusing tool. The exact value and position on the graph is only an indication of relative priority. Individual circumstances could require specific projects to proceed irrespective of their position on the graph.

The actual amount of time that needs to be spent on this Asset Criticality Analysis evaluation exercise is typically five to ten days dependent on facilities included. The assessment culminates with the prioritization of 10 to 15 Asset systems deemed to have the highest impact on overall business performance. Asset reliability improvements can then be focused in those areas that are high in Relative Risk. In addition, an Asset Criticality Number is assigned to the assets associated with the Asset developed in the physical hierarchy. This Asset Criticality Number can be used to establish a criticality index which is used in the EAM/CMMS system to prioritize the execution of maintenance work.

NDS has used this approach at a number of water utilities (e.g. MWRA, Denver Water) with good effect. Sample printouts are included in Appendix C. As mentioned above an additional benefit resulting from this exercise is that the criticality analysis enables the prioritization of maintenance work in the future. The criticality ratings can be stored in the CMMS linked to the asset code and used to prioritize work orders. At the MWRA, NDS created an Access database which is now a valued tool at the MWRA to assist in documenting the ACA results.

The review team for the ACA workshop typically consists of Operations, Maintenance and Technical Specialists identified for their knowledge and experience with facilities. The analysis is built on team consensus of Asset failure impact and frequency. For each Asset system, the most reasonably likely, ***worst case failure scenario*** for the asset system under discussion is analyzed against a utility-customized consequence criteria table for Safety, Environmental, Operational Throughput, and Operational Cost consequences (see example in Appendix C). Each and every consequence score, along with the Asset systems probability / frequency of failure is deemed to be both sensible and defensible by the analysis group.

The Asset Criticality Analysis Tool has been developed to focus Asset reliability improvement activity on business results. Each criticality assessment criterion is assigned a weighting factor, which is used to perform the analysis. The example from the MWRA is included in Appendix C for reference. ***Failure is defined as any time that the Asset performs outside its required standard of performance***. An estimate of Asset reliability is evaluated by assigning a weighting factor indicating the current frequency of failure. An overall "Relative Risk" number relative to all other Assets included in the assessment is calculated for each Asset system line item. The "Relative Risk" number is calculated by multiplying the Total Consequence of Failure by the Frequency/Probability of Failure.

The assessment identifies the following information for each Asset line item:

**A - Total Consequences Number:** This number is the sum of the consequence of failure scores under Safety, Environmental Impact, Throughput, and Operating Cost (for the worst case failure scenario). When compared to the other Assets in the survey it provides an indication of which Assets will have the greatest impact on business performance when they fail. It does not take into account how reliable the Asset is with the current design and practices.

**B - Frequency / Probability of Failure Factor:** This number is an index which indicates the relative probability of failure.

**C - Asset Criticality Number:** This number is between 1 and 9 (9 being most critical). It is calculated using the consequence ratings in the assessment according to an agreed conversion logic. The Asset Criticality Number is used to assist in prioritizing the scheduling of maintenance work.

**D - Relative Risk Number:** The Relative Risk Number (D) is obtained by multiplying the Total Consequence (A) by the Frequency/Probability of Failure weighting Factor (C). This number identifies the Asset that has both significant consequences when it fails and is most likely to fail. The Asset with the highest relative risk number is the Asset most likely to impact business results as a result of its failure. It is used to determine where to focus Asset Reliability Improvement. Asset

Reliability Improvement can include efforts to modify or replace Asset or to change operating or maintenance practices. (Note – the term Total Criticality Number is used in Appendix C for Relative Risk Number.)

## Reliability Program Step 2 – Plan and Schedule RCM Analyses by Criticality Analysis

### *Step 2.1: Level of Effort Assessment for RCM Analysis Planning on Critical Systems*

This task involves further assessment of the systems that were identified as critical systems during the Asset Criticality Analysis. This assessment is required to plan out level of effort required for the RCM analysis for each critical technical system so that these analyses can be prioritized and then scheduled.

In this task, RCM facilitators prepare assessments or estimates of the duration of the RCM analyses (the number of review meetings to be held and the duration of each meeting) for the asset systems defined as critical in the task above.. Part of the assessment includes determination of the boundaries of each technical system to be analyzed and the assets that will be included in each system analysis. [As a yardstick in estimating level of effort required it is our experience that during an RCM analysis of a technical system, an RCM review group working with an experienced RCM facilitator will process/complete approximately six *Failure modes* per hour and the planning and scheduling will be done accordingly.] The "level of effort" for each analysis is estimated in terms of the number of RCM review group meetings required.

As noted above, the level of effort numbers are then plotted against the relative risk numbers to define the areas of initial focus, which should be directed towards items of high risk, low effort or low cost. Those analyses with the potential to deliver the maximum benefit to the company by mitigating risk are selected to be executed first.

It is reiterated that the Asset Criticality Analysis is a focusing tool. The exact value and position on the graph is only an indication of relative priority. Individual circumstances could require specific projects to proceed irrespective of their position on the graph.

### *Step 2.2: Analysis Scheduling Workshop*

The success of the RCM process lies in the participation of the personnel that know the operation and maintenance of the asset systems (Asset) best. These personnel should be invited to take part in each analysis and it is therefore important to consult with the supervisors and the participants (by name and position) well in advance of the exact time and duration of each analysis. It is also important to notify the manager with overall responsibility for the system or asset (who will be responsible for the review and audit that is performed on completion of each analysis) of the analysis program. Detailed planning and scheduling will allow the utility to distribute the workload and schedule personnel involvement in the RCM review groups and audits.

Therefore once priority has been established by the ACA process above, the utility should conduct an Analysis Scheduling workshop to schedule the analyses in detail, utilizing the information from the Asset Criticality Assessment performed in the steps above. The schedule must identify the participants for each analysis and determine the logistics for each project (Meeting dates, meeting times and duration, venues, Asset required etc). The technical information required to perform each analysis will be identified and availability of the same will be discussed. The constraints and assumptions of each analysis will be discussed and formalized.

Once the number of meetings for the asset under review has been estimated it is essential that each meeting is scheduled. The scheduling of each meeting needs to take into consideration the availability of each review group member. The availability of each review group member is determined by his/her commitments such as vacations, training, meetings etc. Once a date, a time and a venue for each meeting has been determined it needs to be scheduled and issued to each group member, his/her supervisor and the RCM project manager. This document should become the formal contract between team members and management to ensure the meetings take place.

Meetings scheduled on an ad-hoc basis leaves the analysis completion open-ended and often results in the inability to get all the team members to the meetings. This becomes inefficient in terms of resource utilization and often extends the number of meetings required to complete the analysis.

Over and above the meeting schedule, the facilitator should identify dates to complete the documentation, the technical audit and the formal hand over of the documentation for the management audit. This allows the development of a formal project plan for the application of the RCM process to the asset under review.

## COSTS AND BENEFITS OF IMPLEMENTING AN RCM PROGRAM (QUANTITATIVE AND QUALITATIVE)

The costs of (1) initiating and then (2) rolling out an RCM program as a large-scale application are as follows:

### Costs for Pilot Site and Program Initiation

The pilot project and RCM program initiation can be completed for approximately $65,000 to $110,000 in external costs and using the internal resources as shown in Tables 4.1 and 4.2.

### Costs for Large-Scale Application on an Annual Basis

The resource usage for large-scale application shown below is based on doing 15 “new” analyses a year (i.e. of different types of asset systems). Use of “templating” would mean coverage of a broad swathe of assets, as many asset systems are replicated in a treatment plant. (Table 5.2)

In essence the large-scale application would utilize 2.5 FTE’s on a full-time basis plus approximately 450 days of other personnel’s time for review groups. Experience would indicate that the critical systems of a large treatment plant would be completed within 1 to 2 years at this level of effort.

Smaller utilities may prefer to look at an approach using contract facilitation – where external resources are used to perform the facilitator role. The recommendation is to still proceed with the review group concept – merely to use an external facilitator where internal resource availability may be tight.

**Table 5.2**
**Costs for large-scale application on an annual basis**

| Item | Qty per RCM analysis | # of RCM analyses | Estimated quantity - annually | Notes |
|---|---|---|---|---|
| Review groups for each analysis | 15 person-days | 30 | 450 person days | 4-5 people for 5-15 meetings for each analysis = 15 man-days per analysis |
| Project Management | 0.5 FTE | | 110 person days | |
| Facilitator | 2 FTE | | 440 person days | Facilitate and write-up 15 analyses per year |
| Project Sponsor | 4 hours per month | | | |
| Steering Committee | 2 hours per month | | | 2 Steering Committee meetings |
| Audit of analysis results | 1-2 days | | 30-40 days | Manager responsible for asset systems analyzed |

## BENEFITS OF RCM PROGRAMS

Effective application of RCM following the approach recommended in this chapter drives multiple benefits. Some of these are tangible (e.g. improved reliability, reduced life-cycle maintenance cost, extended asset life, reduced risk, improved environmental and safety record) while others are more intangible or qualitative (e.g. improved team-work and morale) while other improvements are essentially protective or risk-avoiding (e.g. improved legal position in event of problems occurring).

### Areas Where RCM Drives Benefits

#### *Greater Safety and Environmental Integrity*

The RCM process explicitly considers safety and environmental consequences in the decision tree used to select failure management options (Question 5 of the seven questions or steps stipulated in the SAE standard) –prior to considering operational consequences. The decision tree steps ensure that failure events with safety or environmental consequences are managed to a risk level that is acceptable to the organization. By integrating safety into the mainstream of maintenance decision-making, RCM also improves attitudes to safety.

#### *Improved Operating Performance (Output, Product Quality and Customer Service)*

RCM provides a robust scientifically sound process for deciding which form of maintenance is most suitable in every situation. By doing so, it helps ensure that only the most effective forms of maintenance are chosen for each asset, and that suitable action is taken in cases where maintenance cannot help. This optimized and tightly focused maintenance effort leads to quantum jumps in the performance of existing assets and enables new assets to rapidly ramp up to smooth operation rapidly.

RCM was developed to help airlines draw up maintenance programs for new types of aircraft before they enter service. As a result, it is an ideal way to develop such programs for new assets, especially complex equipment for which no historical information is available. This saves much of the trial and error that is so often part of the development of new maintenance programs – trial that is time-consuming and frustrating, and error that can be very costly and dangerous.

### *Greater Maintenance Cost-Effectiveness*

RCM provides a process to select maintenance activities that have most effect on the performance of the plant. This helps to ensure that everything spent on maintenance is spent where it will do the most good i.e. costs are optimized over the asset life-cycle. As RCM establishes maintenance programs from the failure event level of detail it also provides the input for maintenance simulation programs that simulate asset performance and maintenance cost using Monte Carlo simulation to thoroughly analyze the results of selecting different maintenance concepts and program options over the full asset life-cycle.

In addition, if RCM is correctly applied to existing maintenance systems, it reduces the amount of routine [fixed-interval] work (in other words, maintenance tasks to be undertaken on a cyclic basis) issued in each period, usually by 40% to 70%. On the other hand, if RCM is used to develop a new maintenance program for new assets, the resulting scheduled workload is much lower than if the program is developed by traditional methods.

### *Provides an Optimized Specification for Predictive Technology Tasks to Support Condition-Based Maintenance Where Appropriate*

The RCM process utilizes a logical decision tree analyzes the feasible failure modes of an asset and selects the appropriate failure management policy. Thus the RCM process determines where condition-based maintenance is the appropriate failure management methodology (i.e. is feasible and cost-effective) for specific failure modes and specifies the appropriate condition-monitoring task. [Outputs of the RCM process include an optimized specification for fixed-interval tasks including condition-monitoring tasks and any associated predictive technology.]

Implementation of an effective condition-monitoring program (via the fixed-interval preventive program, operator rounds and on-line, in situ condition monitoring instrumentation) in support of the condition-based maintenance process enables the utility to monitor asset health and to prevent the consequences of applicable failure modes while maximizing the life of the asset (Figure 5.2).

### *A Comprehensive Database Defining the Maintenance Program*

An RCM review ends with a comprehensive and fully documented record of the maintenance requirements of all the significant assets used by the organization. This makes it possible to adapt to changing circumstances (such as changing shift patterns or new technology) without having to reconsider all maintenance policies from scratch. It also enables equipment users to demonstrate that their maintenance programs are built on rational foundations (the audit trail

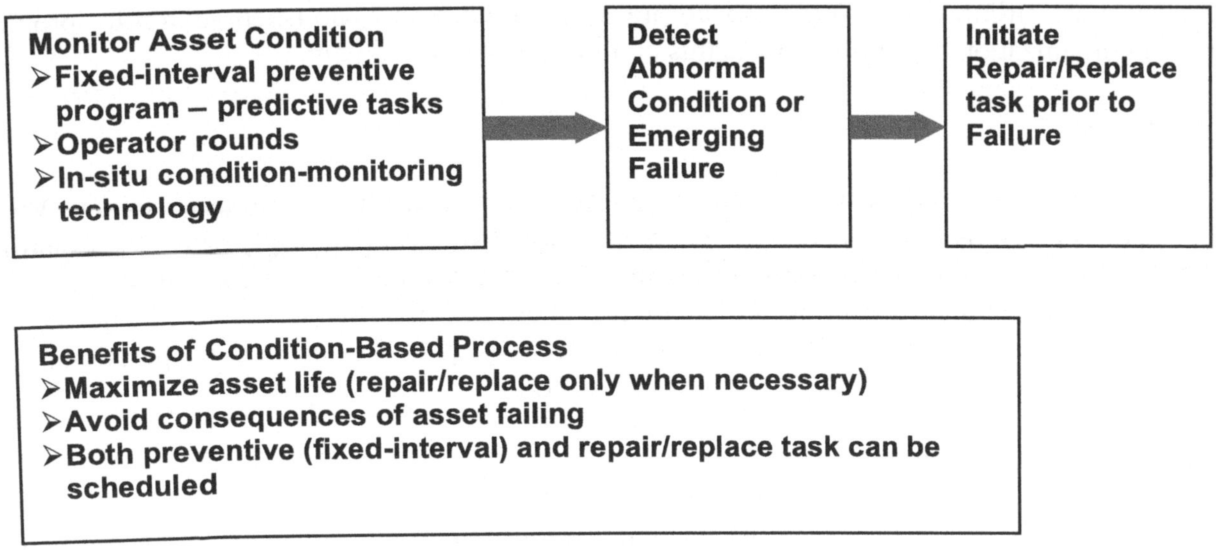


**Figure 5.2 Condition-based maintenance process**

required by more and more regulators). Finally, the information stored on RCM worksheets reduces the effects of staff turnover with its attendant loss of experience and expertise.

An RCM review of the maintenance requirements of each asset also provides a clear view of the skills required to maintain each asset, and for deciding what MRO spares should be held in stock.

### *Optimized MRO Spares Holdings*

A simple extension to the RCM process to develop MRO spares stock holding enables utilities to establish optimized spares holdings on a reliability-based logic. This provides a consistent model for determining MRO holdings and typically reduces inventory while improving MTTR.

### *Greater Motivation of Individuals*

RCM provides greater motivation and improved morale especially for people who are involved in the review process. This is accompanied by much wider 'ownership' of maintenance problems and their solutions. It also means that solutions are more likely to endure.

### *Better Teamwork*

RCM provides a common, easily understood technical language for everyone who has anything to do with maintenance. This gives maintenance, engineering and operations people a better understanding of what maintenance can (and cannot) achieve and what must be done to achieve it. Adhering to the Change Management policies advocated in this chapter ensures that these modern concepts are spread throughout the organization and over time will change the culture of the organization to a reliability-based, proactive model. The differences in paradigm and culture are striking when one visits an organization with an established reliability program versus an organization that has not made this commitment or investment.

### *Reduced Risk*

The RCM program provides a process to manage risk resulting from asset failure. When applied correctly across the asset base of an organization it will systematically move the organization to managing failure and performing maintenance in a manner consistent with its policy of risk (and will ensure that such a policy is set up if it does not exist). In today's world where safety and environmental failures are not tolerated and budget/cost pressures mean that operational failures must be avoided, this is a very important objective.

### *Reduced Insurance Costs*

Organizations which manage risk and can demonstrate to insurers in black and white how they are doing this are able to secure preferential rates. This can provide significant cash savings.

### *Improved Financing – Bond Rates*

Similar logic applies in considering financing costs for major new assets.

### *GASB 34 and Asset Management*

The RCM process and data are congruent with the goals and required approaches for strategic Asset Management and the GASB 34 standard (General Accounting Standards Board) which places new emphasis on measuring and maintaining asset condition. Asset management takes a life-cycle view of assets and their CIP and O and M costs. Utilizing RCM provides an optimal maintenance program to support this life-cycle view. It must be noted that an effective and well-implemented EAM / CMS system is crucial to realize the results of RCM and support the asset management concepts and GASB 34 models. The effects of the GASB34 rules are considered in the RCM decision tree.

## Quantified Benefits

When utilities consider initiating, RCM, management will certainly ask the questions "What is the cost-benefit relationship? or "What is the Payback Period?"

The answers to these questions vary from organization to organization and from asset system to asset system. The evaluations shown in Appendices H and I demonstrate very attractive cost-benefit results and payback periods. However the results of the survey are more ambivalent.

Experience of the authors and within the Aladon network* indicate that quantified cost-benefit and payback will be highly favorable for asset-intensive organizations which are currently

---

* The Aladon network is a world-wide network of RCM consultants that licenses the RCM2 methodology from Aladon Inc, (www.aladon.com) the RCM technology firm founded by John Moubray – one of the senior figures in the industrialization of RCM - and the developers of the RCM2 methodology. This network has implemented RCM in some 2,000 suites worldwide, including many water utilities.

operating with maintenance programs developed on traditional methods (OEM recommendations, gut feel, experience etc) or without any formalized maintenance program and that have not deployed RCM in a strategic, enterprise program utilizing the approach recommended in this chapter.

RCM can usually be cost-justified on the quantitative results in terms of improved operating performance, reduced maintenance costs and reduced costs of failures with relatively short pay-back periods. When one considers that the benefits will be experienced over the remaining life of existing assets the overall cost-benefit is usually highly favorable.

It is also stressed that RCM provides important benefits in areas that are not related solely to cost – particularly in the areas of safety and environmental compliance.

## RECOMMENDED STEP-BY-STEP PROCEDURE FOR IMPLEMENTING AN RCM PROGRAM IDENTIFYING LIKELY RESOURCE NEEDS

The recommended steps for implementing RCM are listed below with estimated durations and then shown Gantt chart format below the listing (Table 5.3 and Figure 5.3).

This reflects a two stage approach as recommended earlier in this chapter. Stage 1 is an RCM Pilot Project followed by Stage 2 – which is the Large-Scale Application of RCM.

Detail on both Stages is provided in the following text.

### Stage 1 - A Systematic Approach to RCM Pilot Projects

One of the recommendations given earlier this chapter in terms of approach is that utilities proceed in stages, using a **pilot project** or **proof-of-concept** (Stage 1) to begin the program and then moving to a large-scale roll-out (Stage 2) of the program. The rationale for this is that the RCM process itself takes time and effort. It also embodies a great deal of new thinking, much of which is surprisingly counter-intuitive. As a result, it is wise for organizations new to this process to approach it in stages. This staged approach is built around the concept of pilot projects designed to enable the organization to learn:

- Exactly what RCM achieves in the water utilities business environment, and how quickly it achieves it. This in turn will enable utilities to decide, before committing to the large-scale application of this process, where the return justifies the investment and to develop a feel for the process and how to best proceed in their own environment.
- How to apply RCM correctly and safely from a purely technical point of view.
- What sort of people (skills) need to be involved directly in its application.

The above summary suggests a sensible way forward would be to carry out a pilot RCM project on two or three assets at two separate water utilities. The two pilot projects performed as part of this research project describe the pilot project concept as occurring in the real world in extensive detail (Appendixes F and G).

As noted earlier under the recommendations to use Review Groups, RCM yields the best results if it is applied by groups of operations and maintenance people working under the direction of RCM facilitators, rather than by single analysts working on their own. Furthermore, the outputs of the RCM process are much more likely to yield substantial and enduring business benefits if the potential benefits to bottom line performance justify the investment.

**Table 5.3**
**List of recommended steps for implementing RC**

| Phase | WBS | Task description | Phase |
|---|---|---|---|
| **(1) Proof-of-concept or Pilot Project Phase** | | | |
| **RCM Pilot Project** | 1 | RCM Implementation Methodology | 304.3d |
| | 1.1 | Introductory presentation to management and stakeholders by RCM consultant | 0.5d |
| | 1.2 | Stage 1 - RCM Pilot Project | 53.3d |
| | 1.2.1 | Discussion of possible pilot project with consultant - after intro. presentation | 1d |
| | 1.2.2 | Assign utility Sponsor for RCM Program | 1d |
| | 1.2.3 | Request proposal from RCM consultants for RCM Pilot Project Training and Services | 5d |
| | 1.2.4 | Award contract to RCM consultants for RCM Pilot Project Training and Services | 5d |
| | 1.2.5 | 3 day introductory course for people involved with pilot project | 3d |
| | 1.2.6 | Identify-Plan-Schedule Pilot Project | 13.5d |
| | 1.2.6.1 | Develop specific measurable objectives for pilot project | 0.5d |
| | 1.2.6.2 | Perform criticality analysis on asset base or sub-set of asset base | 10d |
| | 1.2.6.3 | "Select asset system(s) and define scope, number of RCM analyses" | 0.5d |
| | 1.2.6.4 | Assign utility Project Manager for RCM pilot project | 0.5d |
| | 1.2.6.5 | Define review group members for each RCM analysis in Pilot | 0.5d |
| | 1.2.6.6 | Schedule review group meetings and define location | 0.5d |
| | 1.2.6.7 | "Communicate with review group members, supervisors and managers" | 0.5d |
| | 1.2.6.8 | Procure software license or agree to use consultant software | 1d |

(continued)

**Table 5.3 (Continued)**

| Phase | WBS | Task description | Phase |
|---|---|---|---|
| RCM Pilot Project | 1.2.7 | Execute Pilot Project (using external facilitator and rented software) | 53d |
| | 1.2.7.1 | 3 day RCM training for any review group members not trained | 3d |
| | 1.2.7.2 | Pilot Project RCM Analysis 1 | 28d |
| | 1.2.7.2.1 | Conduct review group meetings (5-15 3 hour meetings per analysis) | 15d |
| | 1.2.7.2.2 | Write up analysis report | 3d |
| | 1.2.7.2.3 | Audit and implement Analysis Recommendations | 10d |
| | 1.2.7.3 | Pilot Project RCM Analysis 2 | 28d |
| | 1.2.7.3.1 | Conduct review group meetings (5-15 3 hour meetings per analysis) | 15d |
| | 1.2.7.3.2 | Write up analysis report | 3d |
| | 1.2.7.3.3 | Audit and implement Analysis Recommendations | 10d |
| | 1.2.7.4 | Pilot Project RCM Analysis n | 28d |
| | 1.2.7.4.1 | Conduct review group meetings (5-15 3 hour meetings per analysis) | 15d |
| | 1.2.7.4.2 | Write up analysis report | 3d |
| | 1.2.7.4.3 | Audit and implement Analysis Recommendations | 10d |
| | 1.2.7.5 | Prepare management presentation | 28d |
| | 1.2.7.5.1 | Measure short-term results of pilot project | 15d |
| | 1.2.7.5.2 | Compile into PowerPoint presentation | 3d |
| | 1.2.8 | Analyze Pilot Project—Present results to management | 10d |
| | 1.2.8.1 | Present results achieved and feedback on RCM process | 5d |
| | 1.2.8.2 | Include possible next steps as part of presentation | 3d |

(continued)

**Table 5.3 (Continued)**

| Phase | WBS | Task description | Phase |
|---|---|---|---|
| **(2) Large-scale Implementation Phase** | | | |
| RCM Large-Scale Implementation | 1.3 | Stage 2 - Large-Scale Application of RCM [RCM LSA] | 251d |
| | 1.3.1 | Identify Stage 2 RCM LSA Scope | 28d |
| | 1.3.1.1 | Appoint Project Manager for Large-scale Application (Roll-out) of RCM | 1d |
| | 1.3.1.2 | Appoint Steering Committee for Large-scale application of RCM | 1d |
| | 1.3.1.3 | Define objectives of large-scale application of RCM | 5d |
| | 1.3.1.4 | Prepare high level approach and project plan | 5d |
| | 1.3.1.5 | Prepare budget | 5d |
| | 1.3.1.6 | Prepare business case | 10d |
| | 1.3.1.7 | Present Business Case to Management for Approval | 1d |
| | 1.3.2 | Plan-Schedule RCM LSA | 40d |
| | 1.3.2.1 | Engage External Services for RCM Training and Guidance | 25d |
| | 1.3.2.1.1 | Prepare RFP for external services for RCM Training and Guidance-Mentoring | 10d |
| | 1.3.2.1.2 | Contract firm for RCM Training and Services for Large-scale Application | 10d |
| | 1.3.2.1.3 | Revise or add detail to criticality analysis as required | 5d |
| | 1.3.2.2 | Perform analysis planning workshop using critically analysis | 5d |
| | 1.3.2.3 | Select internal facilitators | 5d |
| | 1.3.2.4 | Train internal facilitators - 3 day RCM course plus 10 day RCM Facilitator Course | 15d |
| | 1.3.2.5 | Procure RCM software | 10d |
| | 1.3.2.6 | Load RCM software on network | 2d |
| | 1.3.2.7 | Set up users and access rights to RCM software | 3d |
| | 1.3.2.8 | Schedule RCM analyses | 5d |
| | 1.3.2.9 | Organize meeting facilities for review groups as per RCM analysis schedule | 2d |
| | 1.3.2.10 | Select Review Group Members | 5d |
| | 1.3.2.11 | "Communicate with review group members, supervisors and managers" | 5d |
| | 1.3.2.12 | Plan for RCM 3 day training course for ALL Review Group members | 5d |

(continued)

**Table 5.3 (Continued)**

| Phase | WBS | Task description | Phase |
|---|---|---|---|
| RCM Large-Scale Implementation | 1.3.3 | Execute RCM LSA | 185d |
| | 1.3.3.1 | Train RCM Review Group Members on 3 day courses as required | 150d |
| | 1.3.3.2 | Specify items to be documented for before-after comparison | 10d |
| | 1.3.3.3 | Develop processes and procedures for RCM / Maintenance Excellence | 60d |
| | 1.3.3.3.1 | Review Group Process | 20d |
| | 1.3.3.3.2 | RCM Analysis Report Format and Contents | 20d |
| | 1.3.3.3.3 | "Review, Management Audit and Approval/Sign-off" | 20d |
| | 1.3.3.3.4 | RCM involvement in new asset/ modification processes | 20d |
| | 1.3.3.4 | Perform RCM analyses as per schedule | 150d |
| | 1.3.3.5 | RCM consultant to mentor and guide utility facilitators | 150d |
| | 1.3.3.6 | Audit and Implement results as each analysis is completed | 150d |
| | 1.3.3.7 | Implement or Remediate/ Improve EAM/CMMS system as required to execute RCM maintenance program | 100d |
| | 1.3.3.7.1 | Update asset configuration in EAM/CMMS to agree with real world | 60d |
| | 1.3.3.7.2 | Formalize core maintenance processes and train workforce | 80d |
| | 1.3.3.7.3 | Ensure key maintenance reports are created and distributed | 40d |
| | 1.3.3.8 | Use Criticality Number on Assets to prioritize maintenance work | 20d |
| | 1.3.3.9 | Implement intranet to publicize results | 20d |
| | 1.3.3.10 | Set up fixed-interval tasks to review RCM analyses every 18 months | 5d |
| | 1.3.4 | Control-Analyze RCM LSA | 150d |
| | 1.3.4.1 | Report on variables measuring achievement of objectives | 150d |
| | 1.3.5 | Program Management Office For RCM LSA | 250d |
| | 1.3.5.1 | Steering Committee meets monthly | 250d |
| | 1.3.5.1.1 | Monitor and adjust project progress | 250d |

| ID | | Task Name | Duration | Start |
|---|---|---|---|---|
| 1 | | **RCM IMPLEMENTATION METHODOLOGY** | **304.3 days?** | **Mon 9/5/05** |
| 2 | | Introductory presentation to management and stakeholders by RCM consultant | 0.5 days | Mon 9/5/05 |
| 3 | | **STAGE 1 - RCM PILOT PROJECT** | **53.3 days?** | **Mon 9/5/05** |
| 4 | | Discussion of possible pilot project with consultant - after intro. presentation | 1 day | Mon 9/5/05 |
| 5 | | Assign utility Sponsor for RCM Program | 1 day | Tue 9/6/05 |
| 6 | | Request proposal from RCM consultants for RCM Pilot Project Training and Services | 5 days | Wed 9/7/05 |
| 7 | | Award contract to RCM consultants for RCM Pilot Project Training and Services | 5 days | Wed 9/14/05 |
| 8 | | 3 day introductory course for people involved with pilot project | 3 days | Wed 9/28/05 |
| 9 | | **IDENTIFY-PLAN-SCHEDULE PILOT PROJECT** | **13.5 days?** | **Mon 10/3/05** |
| 10 | | Develop specific measurable objectives for pilot project | 0.5 days | Mon 10/3/05 |
| 11 | | Perform criticality analysis on asset base or sub-set of asset base | 10 days | Tue 10/4/05 |
| 12 | | Select asset system(s) and define scope, number of RCM analyses | 0.5 days | Tue 10/18/05 |
| 13 | | Assign utility Project Manager for RCM pilot project | 0.5 days | Mon 10/3/05 |
| 14 | | Define review group members for each RCM analysis in Pilot | 0.5 days | Tue 10/18/05 |
| 15 | | Schedule review group meetings and define location | 0.5 days | Wed 10/19/05 |
| 16 | | Communicate with review group members, supervisors and managers | 0.5 days | Wed 10/19/05 |
| 17 | | Procure software licence or agree to use consultant software | 1 day? | Thu 10/20/05 |
| 18 | | **EXECUTE PILOT PROJECT (using external facilitator and rented software)** | **53 days** | **Mon 9/5/05** |
| 19 | | 3 day RCM training for any review group members not trained | 3 days | Fri 10/21/05 |
| 20 | | **Pilot Project RCM Analysis 1** | **28 days** | **Mon 9/5/05** |
| 21 | | Conduct review group meetings (5-15 3 hour meetings per analysis) | 15 days | Mon 9/5/05 |
| 22 | | Write up analysis report | 3 days | Mon 9/26/05 |
| 23 | | Audit and implement Analysis Recommendations | 10 days | Thu 9/29/05 |
| 24 | | **Pilot Project RCM Analysis 2** | **28 days** | **Mon 9/19/05** |
| 25 | | Conduct review group meetings (5-15 3 hour meetings per analysis) | 15 days | Mon 9/19/05 |
| 26 | | Write up analysis report | 3 days | Mon 10/10/05 |
| 27 | | Audit and implement Analysis Recommendations | 10 days | Thu 10/13/05 |
| 28 | | **Pilot Project RCM Analysis n** | **28 days** | **Mon 10/3/05** |
| 29 | | Conduct review group meetings (5-15 3 hour meetings per analysis) | 15 days | Mon 10/3/05 |

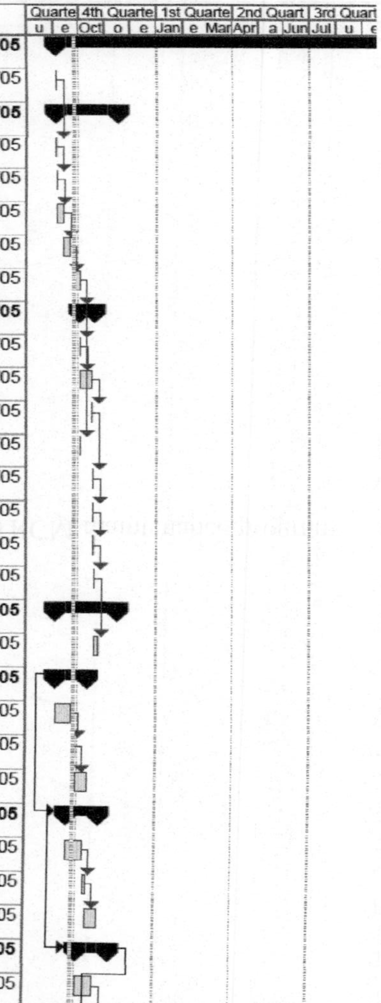


Project: Recommended steps for RCM
Date: Sun 9/25/05

Task · Split · Progress · Milestone · Summary · Project Summary · External Tasks · External Milestone · Deadline

Page 1

**Figure 5.3 Gantt chart showing recommended steps for RCM program**

(continued)

| ID | Task Name | Duration | Start |
|---|---|---|---|
| 30 | Write up analysis report | 3 days | Mon 10/24/05 |
| 31 | Audit and implement Analysis Recommendations | 10 days | Thu 10/27/05 |
| 32 | **Prepare management presentation** | **5 days** | **Thu 11/10/05** |
| 33 | Measure short-term results of pilot project | 3 days | Thu 11/10/05 |
| 34 | Compile into PowerPoint presentation | 2 days | Tue 11/15/05 |
| 35 | **ANALYZE PILOT PROJECT Present results to management** | **0.3 days** | **Thu 11/17/05** |
| 36 | Present results achieved and feedback on RCM process | 0.3 days | Thu 11/17/05 |
| 37 | Include possible next steps as part of presentation | 0.2 days | Thu 11/17/05 |
| 38 | **STAGE 2 - LARGE-SCALE APPLICATION OF RCM [RCM LSA]** | **251 days** | **Thu 11/17/05** |
| 39 | **IDENTIFY STAGE 2 RCM LSA SCOPE** | **28 days** | **Thu 11/17/05** |
| 40 | Appoint Project Manager for Large-scale Application (Roll-out) of RCM | 1 day | Thu 11/17/05 |
| 41 | Appoint Steering Committee for Large-scale application of RCM | 1 day | Fri 11/18/05 |
| 42 | Define objectives of large-scale application of RCM | 5 days | Mon 11/21/05 |
| 43 | Prepare high level approach and project plan | 5 days | Mon 11/28/05 |
| 44 | Prepare budget | 5 days | Mon 12/5/05 |
| 45 | Prepare business case | 10 days | Mon 12/12/05 |
| 46 | Present Business Case to Management for Approval | 1 day | Mon 12/26/05 |
| 47 | **PLAN-SCHEDULE RCM LSA** | **40 days** | **Mon 12/26/05** |
| 48 | **Engage External Services for RCM Training and Guidance** | **25 days** | **Mon 12/26/05** |
| 49 | Prepare RFP for external services for RCM Training and Guidance-Mentoring | 10 days | Mon 12/26/05 |
| 50 | Contract firm for RCM Training and Services for Large-scale Application | 10 days | Mon 1/9/06 |
| 51 | Revise or add detail to criticality analysis as required | 5 days | Mon 1/23/06 |
| 52 | Perform analysis planning workshop using critically analysis | 5 days | Mon 12/26/05 |
| 53 | Select internal facilitators | 5 days | Mon 1/23/06 |
| 54 | Train internal facilitators - 3 day RCM course plus 10 day RCM Facilitator Course | 15 days | Mon 1/30/06 |
| 55 | Procure RCM software | 10 days | Mon 1/23/06 |
| 56 | Load RCM software on network | 2 days | Mon 2/6/06 |
| 57 | Set up users and access rights to RCM software | 3 days | Wed 2/8/06 |
| 58 | Schedule RCM analyses | 5 days | Mon 1/2/06 |

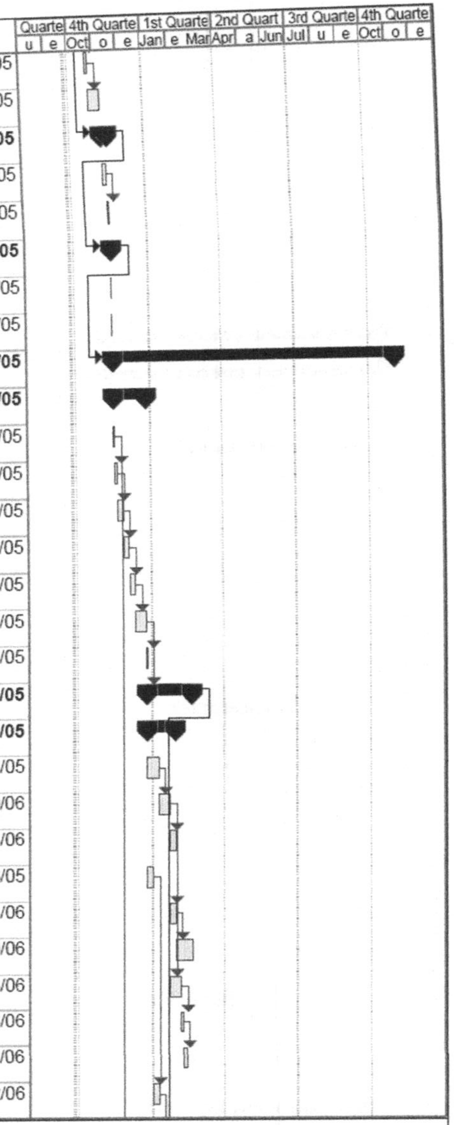


Project: Recommended steps for RCM
Date: Sun 9/25/05

Task | Split | Progress | Milestone | Summary | Project Summary | External Tasks | External Milestone | Deadline

Page 2

Figure 5.3 (Continued)

(continued)

| ID | Task Name | Duration | Start |
|---|---|---|---|
| 59 | Organize meeting facilities for review groups as per RCM analysis schedule | 2 days | Mon 1/9/06 |
| 60 | Select Review Group Members | 5 days | Mon 1/9/06 |
| 61 | Communicate with review group members, supervisors and managers | 5 days | Mon 1/16/06 |
| 62 | Plan for RCM 3 day training course for ALL Review Group members | 5 days | Mon 1/16/06 |
| 63 | **EXECUTE RCM LSA** | **185 days** | **Mon 1/23/06** |
| 64 | Train RCM Review Group Members on 3 day courses as required | 150 days | Mon 1/23/06 |
| 65 | Specify items to be documented for before-after comparison | 10 days | Mon 1/23/06 |
| 66 | **Develop processes and procedures for RCM / Maintenance Excellence** | **60 days** | **Mon 1/23/06** |
| 67 | Review Group Process | 20 days | Mon 1/23/06 |
| 68 | RCM Analysis Report Format and Contents | 20 days | Mon 2/6/06 |
| 69 | Review, Management Audit and Approval/Sign-off | 20 days | Mon 2/20/06 |
| 70 | RCM involvement in new asset/ modification processes | 20 days | Mon 3/20/06 |
| 71 | Perform RCM analyses as per schedule | 150 days | Mon 1/30/06 |
| 72 | RCM consultant to mentor and guide utility facilitators | 150 days | Mon 3/13/06 |
| 73 | Audit and Implement results as each analysis is completed | 150 days | Mon 2/20/06 |
| 74 | **Implement or Remediate/ Improve EAM/CMMS system as required to execute R(** | **100 days** | **Mon 2/20/06** |
| 75 | Update asset configuration in EAM/CMMS to agree with real world | 60 days | Mon 2/20/06 |
| 76 | Formalize core maintenance processes and train workforce | 80 days | Mon 2/20/06 |
| 77 | Ensure key maintenance reports are created and distributed | 40 days | Mon 5/15/06 |
| 78 | Use Criticality Number on Assets to prioritize maintenance work | 20 days | Mon 6/12/06 |
| 79 | Implement intranet to publicize results | 20 days | Mon 1/23/06 |
| 80 | Set up fixed-interval tasks to review RCM analyses every 18 months | 5 days | Mon 8/21/06 |
| 81 | **CONTROL-ANALYZE RCM LSA** | **150 days** | **Mon 2/27/06** |
| 82 | Report on variables measuring achievement of objectives | 150 days | Mon 2/27/06 |
| 83 | **PROGRAM MANAGEMENT OFFICE FOR RCM LSA** | **250 days** | **Fri 11/18/05** |
| 84 | **Steering Committee meets monthly** | **250 days** | **Fri 11/18/05** |
| 85 | Monitor and adjust project progress | 250 days | Fri 11/18/05 |

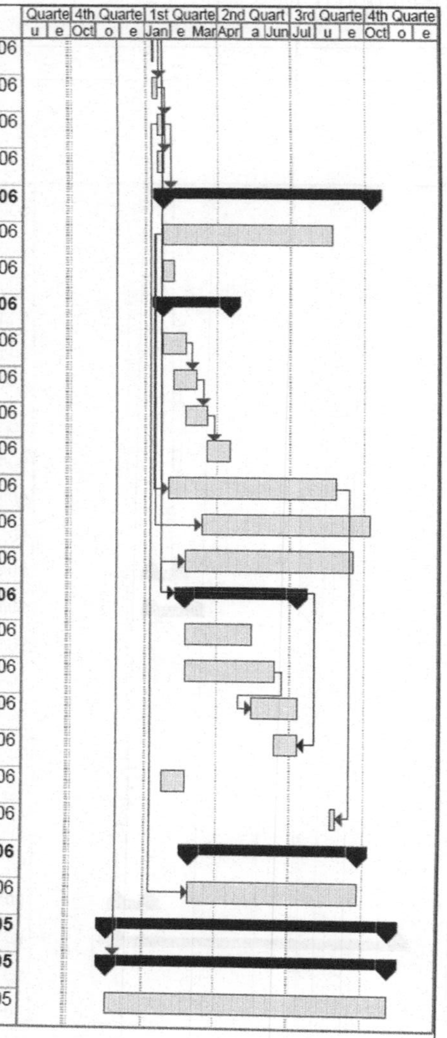


Project: Recommended steps for RCM
Date: Sun 9/25/05

Task | Split | Progress | Milestone | Summary | Project Summary | External Tasks | External Milestone | Deadline

Page 3

Figure 5.3 (Continued)

The pilot projects performed at Denver Water and at Veolia Water Indianapolis have utilized the RCM2 approach developed by Aladon Inc. Over the past 20 years the Aladon network (of which NDS is a licensee) has developed a suite of training and mentoring services that transfer the technology of RCM to end-users as efficiently as possible. This suite consists of four key elements:

- A basic three-day introductory training course for RCM review group members and facilitators (more than 60,000 people have attended this course around the world)
- Further formal training and on-site mentoring for RCM facilitators (the Aladon network has now trained more than 2,500 RCM facilitators)
- Advice and guidance on the management of RCM projects (the Aladon network has helped companies to apply RCM on more than 2,000 sites)
- RCM software (the Aladon RCM software package is operating successfully in more than 400 locations).

Pilot RCM projects are most likely to succeed if they are carried out in four stages, as follows:

1. *Introductory course:* Arrange a three-day on-site RCM course for a cross-section of maintenance and operations managers and union leaders from 2 or 3 candidate pilot areas
2. *Conduct an equipment criticality analysis:* This entails first of all ranking the equipment by "relative risk" they pose in preventing the attainment of business goals and secondly, prioritizing the order in which critical assets should be addressed by RCM
3. *Prepare a project plan* (in consultation with those who have been through steps '1' and '2'): The project planning step entails establishing detailed objectives for the pilot projects (in other words, deciding exactly how "success" is to be judged) and deciding what equipment is to be analyzed, who is to perform the analysis and how long it will take.
4. *Do the pilot projects* as planned.

When the pilot projects have been completed, the organization is in a position to roll out RCM in areas where the investment of time, effort and money looks worthwhile, based on an evaluation of what RCM achieves when applied to its own assets by its own people. The four stages of a pilot project are discussed in detail in the sections below of this proposal.

## Introductory Course

As stressed in the recommendations on Change Management, RCM embodies a great deal of new thinking, much of which is contrary to what many maintenance people believe - or may have even been taught. As a result, one of the most critical success factors in the application of RCM is not only to apply the structured RCM methodology, but also to change the way people think. We have found that if too little emphasis is placed on this aspect of the change management process, the organization either applies RCM incorrectly or it simply rejects any new maintenance regime that emerges from the RCM process (often both), and the associated investment is wasted.

For this reason, we have found that any pilot application of RCM should start with an introductory course. This course is not only intended for key participants in the process, but also for people (usually managers and supervisors) who are going to be asked to release people to participate in the process and to apply the results. In this context, the objectives of our three-day introductory course are as follows:

- To provide review group members with enough of an understanding of the basic principles of RCM for them to be able to participate meaningfully in the process
- To provide entry level training for RCM facilitators
- To ensure that first line supervisors have a sufficient grasp of the objectives and technical content of RCM for them to be able to participate sensibly in planning an RCM project, and subsequently in helping to ensure that it is completed as planned.

**Equipment Criticality**

[Note: This task may be performed as part of the pilot project or large-scale application depending on resources available for the pilot]

The theory and best practice around asset criticality has been discussed at length earlier in this chapter – it is summarized here again as part of the step-by-step process. Equipment reliability improvement projects compete with the many other improvement programs requiring funding and people. If a water utility is to be successful at deploying RCM, it is essential that a business case is developed to justify the expenditure of resources. Management must believe that the potential benefits to the company's bottom line performance justify the investment. The challenge is to rise to the opportunity this represents. As a result, it is recommended to use an equipment criticality evaluation tool for this purpose.

The *Asset Criticality Analysis Tool* has two purposes. First, rank equipment by the "relative risk" they pose in preventing the attainment of business goals and secondly, prioritize the order in which critical assets should be addressed by RCM. Achieving this requires consensus between stakeholders on the result. More precisely, the right information has to be collected from the right people. The Asset Criticality Review group should include operations, maintenance and technology members; at the management, supervision and first line level. Collectively they process the knowledge and responsibility to deal with the question:

"Where do we need to focus our equipment reliability improvement efforts to maximize contribution to business goals?"

The next step looks at how difficult it will be to improve the reliability of the critical equipment in terms of the level of effort required. In this case level of effort reflects the number of RCM review group meetings required to analyze the equipment. The level of effort numbers are plotted against the relative risk numbers to define the areas of initial focus, which should be directed towards items of high risk, low effort or low cost. Projects with the potential to deliver the maximum benefit to the company by mitigating the risk are selected to be executed first.

It is noted that the ACA Tool is a focusing tool. The exact value and position on the graph is only an indication of relative priority. Individual circumstance could require specific projects to proceed irrespective of their position on the graph.

The actual amount of time that normally needs to be spent on the asset criticality evaluation exercise would be five to ten days dependent on size of facility/asset base.

### Pilot Project Planning

As with any moderately resource intensive initiative, the next critical success factor is to ensure that the project is planned as thoroughly as possible before work starts. In order to ensure that the plan is realistic and in order to secure a reasonable degree of ownership of the plan, the plan must be prepared in close consultation with the managers and supervisors of the areas affected. In preparing the plan, the following questions need to be answered:

- What equipment will be covered by the project as determined by the criticality analysis?
- To what extent is equipment duplicated?
- How many RCM meetings are needed to analyze each item?
- How many people at each level will be directly involved in review groups?
- How many different review groups are needed? (The answer to this question is governed by the answers to the previous questions)
- How many review groups will each facilitator work with at once (in other words, how many review projects will be in progress at once)?
- What other factors might have a bearing on the manning, the start date and the duration of each analysis (such as vacation and other training commitments of key people, parallel initiatives, shutdowns, shift arrangements, etc.)?
- On what basis are the reviews to be audited, and by whom is this to be done in each plant section?

A second objective of this exercise is to establish the objectives of the pilot projects as clearly as possible, and to agree how management would decide whether or not these had been achieved at the end of each project.

The answers to these questions will enable us to prepare a draft program for the pilot section(s), and to identify what training is needed at each stage. The actual amount of time that needs to be spent on this planning exercise depends on the number and sizes of the areas selected, and hence a firm project plan could only be submitted once the scope of the study has been decided. This having been said, we are usually able to develop a reasonably thorough plan in period of three days, of which two are normally spent on site and one off-site preparing a written report.

### The Pilot Project

#### *RCM Analyses*

The RCM pilot project itself begins with a three day RCM training course for any review group members who have not been trained. Thereafter, the analysis of each system consists of the following steps:

- The maintenance requirements of the equipment in each system are assessed by teams made up of operations and maintenance people who know the equipment best in that section. Each team works under the guidance of an RCM facilitator. We recommend that each group meets for not less than three and not more than four hours on any one day. Meetings can be held on a part-time basis at the rate of one or two per week, or an intensive approach can be adopted, with groups meeting on consecutive days until their analysis is complete. (The approach is usually decided during the project-planning phase.)

- When each RCM analysis has been completed, it is formally audited by the senior managers who have overall responsibility for the assets concerned. Their role is to satisfy themselves that the maintenance programs arising from the analyses are sensible and defensible.
- When the review has been approved, steps are taken to ensure that the tasks are performed as required, as discussed below.
- We recommend that the projects end with a formal presentation to top management. The facilitators or review group members should give this presentation. It should provide a summary of the findings and conclusions of each group, together with an evaluation of the benefits of RCM in relation to the time and effort needed to apply it (with special reference to the specific objectives of each analysis, as outlined earlier).

### *Facilitation*

Pilot projects can be facilitated by external (consultant) or internal facilitators. Use of experienced, external facilitators expedites the pilot projects and gives a sense of productivity using experienced facilitators. If internal facilitators are used, they must be put through a ten day RCM facilitator's course and mentored on their first analyses until they pick up experience. A mixed approach can also be used where some of the pilot project RCM analyses are facilitated by external RCM consultants and some by newly trained internal facilitators mentored by experienced RCM consultants.

In the Pilot projects included as part of this research project, NDS utilized an experienced RCM practitioner to complete the first analysis on both water utilities pilot projects to provide a benchmark to assess how much would be required to roll out RCM if the pilot projects were successful. This is because newly trained facilitators take two or three times as long to complete their first RCM analyses as they take after they have gained some experience.

### *Training*

The further formal training needed to support the pilot project consists of two parts:

- Training for group members who have not been trained.
- A ten-day training course for facilitators if internal facilitators are used

Each of these issues is discussed below.

### *Training for Group Members*

As mentioned in this research report, it is essential that all the people who participate directly in an RCM analysis attend a three-day training course. It may be possible to train all the prospective participants on the introductory course described above, but our experience indicates that the project planning process is likely to reveal that more people will need to be trained. Additional training courses, if necessary, are confirmed during the pilot project planning.

Each three-day course in the research project pilots was delivered on premises provided by the participating water utility for a maximum of sixteen delegates on each course. The utilities provided:

- A room capable of seating the delegates at tables in a "U"-formation
- Two flip charts, an SVGA digital projector and a screen
- Writing materials and stationery for the delegates
- Catering considered to be appropriate for a course of this nature.

NDS provided printed notes for each delegate. The participating utility manager was responsible for ensuring that delegates actually do attend.

### *Facilitator Training*

Aladon has identified 45 distinct skills required by RCM facilitators. The facilitator training process used for the research project pilots is designed to develop the competency of facilitators in all 45 skills to the point that they are able to produce safe, defensible analyses without further support. The competency development process takes place in three stages. The first stage consists of the basic three-day RCM course.

The second stage entails a formal ten-day training course. The facilitators learn how to guide a group in the application of RCM in addition to consolidating their own knowledge of the subject. Under the guidance of our consultant, they work as a group on a series of exercises. They take turns to act as facilitator, and the exercises consist of one standard case study provided by Aladon and others based on items of plant chosen from the site at which the course is held. (These ten-day training courses are usually held at regular intervals on a syndicated basis for facilitators from a number of different client organizations. The venue for the course attended by the research project facilitators was provided by one of the participating water utilities.)

The third stage of facilitator training consists of the on-site mentoring. This entails monitoring the work done by the facilitators in particular. It also entails ensuring that the group members and auditors understand the philosophy and apply it correctly.

It is recommended that experienced RCM consultants provide this support, especially in the early stages of the project. The amount of support needed depends on how thoroughly the facilitators grasp the concepts of RCM, and how well they apply them. Final conclusions can only be drawn in this regard once the group meetings are under way. Experience has shown that on average, facilitators require between ten to twenty five days of this support, the average being about fifteen days per facilitator, spread over their first two to four analyses.

### *RCM Project Management*

An RCM project needs to be managed by one or two designated people who work with the managers of the pilot area to carry out the following:

- Decide who will attend which training courses on which dates (please note that it is essential to keep meticulous records listing who has undergone what type of

RCM training, especially if there is a likelihood that RCM will be taken further in the long term)

- Arrange venues, stationery and catering for the courses
- Ensure that delegates arrive at the designated courses, and take appropriate action if any delegates who fail to arrive
- Arrange a suitably furnished venue for the meetings, and a suitable work area for the facilitators
- Ensure that all group members and their supervisors know when their review meetings are due to take place
- Ensure that the members of each group attend their meetings, and take action in the event of non-attendance
- Ensure that audits are carried out as soon as possible after each review has been completed
- Keep senior management informed on progress, and to assist with the preparation of the presentations to senior management at the end of each project.

***Implementation***

The three principle outcomes of any RCM analysis are recommendations on:

- New or revised maintenance routines
- Revised operating procedures
- Equipment design changes.

In order to consolidate and to sustain any improvements achieved during the RCM analysis, it is essential that steps are taken to implement these recommendations on a formal basis as soon as possible after the review meetings have been completed.

It is noted that the improvements resulting from an RCM project often start to flow right after the first meeting has been held, and not only after all the recommendations have been implemented. In fact, in some cases, nearly all of the improvements are achieved while the analyses are still under way (to the extent that some users see the implementation phase purely as a way of sustaining improvements).

***RCM Software***

The pilot projects must be supported by use of an appropriate RCM software package. Depending on how the pilot project is structured the recommendation is to have the RCM consultants provide a copy of the RCM software to be used for the pilot or for the utility to rent the software during the pilot phase, so it can be evaluated before actually procuring the licenses.

The software used for the Pilot projects completed as part of this project was the Aladon RCM2 Toolkit. This package has been designed to support RCM facilitators by providing a central database for storing and sorting information generated by RCM review groups. It enables them to print out information and decision worksheets in the standard Aladon formats. It also enables users to sort tasks by skill-set, frequency and other criteria to produce work packages, and it can be used to generate a variety of other reports. It runs stand-alone or as a network product in the Windows environment.

**Table 5.4**
**Business functions and role of personnel requiring RCM training**

| Business Function | | Management level / role | | |
|---|---|---|---|---|
| | | Executive | Operational | Tactical |
| | General Management | 1D | | |
| | Maintenance Function | 1D | 3D | 3D |
| | Operations | 1D | 3D | 3D |
| | Engineering | 1D | 3D | 3D |
| | Reliability | 1D | 3D AND 10D | 3D AND 10D |
| | Others | 1D | | |

1D = 1 Day Management Overview on RCM and EAM Best Practice
3D = 3 Day RCM Introduction (Overview) Course – Theory and Case Studies
10D = 10 Day RCM Facilitation Course

## CHANGE MANAGEMENT RECOMMENDATIONS TO MOVE FROM BREAKDOWN REACTIVE APPROACH TO PROGRAM BUILT ON RELIABILITY-CENTERED PRINCIPLES

The Change Management recommendations have been mentioned in some of the sections above and are discussed further in the following section.

### Senior Management Ownership / Sponsorship for the RCM Program

As noted in Critical Success factors under Recommended Approach, real success using RCM will only be achieved as a strategic program driven from the top of the organization. Isolated groups may achieve positive results in analyzing specific asset systems, but structural, systemic change will only be achieved with senior management ownership of and commitment to the program. The reasons for this include the following:

Sustainable reliability programs require concerted change management and training programs which require significant commitments of funds and key people's time. Many of the tenets of RCM require a fundamental shift in the way people think about maintenance. Such systemic change will typically only occur in a there is this commitment from the top of the organization.

### Extensive Training of Stakeholders and RCM Review Group Members

Training and education are critical aspects of the change management process towards maintenance excellence and a proactive, controlled maintenance environment based on an RCM-derived maintenance program. (See Table 5.4.)

Training should be structured and targeted to the level of management and to the detail required. Senior or executive management should receive a one day overview on reliability concepts and what RCM does and other maintenance excellence concepts. Personnel from the other functional departments shown should, over time; all attend a three day course on RCM and be involved in review groups. Training on other aspects of maintenance excellence and effective use of the CMMS/EAM system must not be overlooked.

A strategic training initiative in tandem with the Reliability program achieving positive results can produce a sea change in the organization in terms of understanding modern maintenance concepts, adopting a pro-active approach and believing in a reliability-based maintenance program. The business results from addressing reliability and maintenance excellence in a strategic manner can be quite startling.

**Use of Review Group Concept**

As noted in Critical Success factors under Recommended Approach, the review group approach is a critical component for successful change management. Comprehensive involvement of maintenance, engineering and operations people means that these personnel are:

- well trained in the RCM process and best practice around Reliability
- actively involved with using the process
- jointly "own" the results.

This is a recipe for Change Management success.

**Maintenance Work Managed and Controlled by Effective Maintenance Management System**

One of the requirements of a move to a pro-active controlled environment is that maintenance work must be managed and controlled by a maintenance management system (typically a computerized CMMS or EAM system). If the maintenance workforce works to their own agenda and continues to work reactively, little will change. The disciplines and process must be in place where work is planned and scheduled in the maintenance management system and the maintenance work force works according to this plan and schedule – with exception of urgent breakdown work. This is key – for the benefits of RCM to be realized the fixed-interval work must be completed as scheduled.

Therefore if the CMMS or EAM system is ineffective or does not exist, the utility must fix this and implement and effective maintenance management system to ensure that the right work is executed at the right time.

Implementation of effective EAM/CMMS systems is a subject in its own right and requires effective change management steps of its own. The RCM program will have very little value if the tasks are not scheduled and executed as specified.

**Organization Structure Adjustments**

A move to maintenance excellence and a proactive program built on best principles is accomplished by actions on several fronts. One of these fronts is that the organization must be designed to fit the revised business processes. This does not mean more people – it means using the same (or possibly less) resources effectively. Part of the change management processes is creating and correctly designing these positions, filling them with the most appropriate personnel and training these people how to do their new job.

### *Reliability*

The role of analyzing failure, reliability problems and continuing to optimize Reliability should be handled by a specific person. Most major asset-intensive organizations have people focused on reliability to ensure remedial action. If this is just left to the maintenance line organization it will be lost under the operational pressures of executing work.

### *Planning and Scheduling*

As mentioned earlier, strong planning and scheduling processes supported by a well-implemented EAM/CMMS system are an integral part of the move from reactive maintenance to a controlled proactive environment. The Planning and Scheduling roles MUST be broken out into dedicated roles and staffed with people who have bee trained to do this function.

Planning is defining scope of a task and the resources required; scheduling is assigning the task to a time and crew. Effective Planning and Scheduling of Maintenance work is absolutely crucial to proactive maintenance – if this does not occur maintenance crews typically relapse into the brake-fix mentality. Again part of the change management processes is creating and correctly designing these positions if they do not exist, filling them with the most appropriate personnel and training these people how to do their new job.

### *Project Manager for RCM Program*

Responsibility for managing and driving the RCM program must be assigned to person who is capable of running the program. This role can either be full-time or part-time depending on the size of the utility and intensity of the RCM program.

### *Facilitators*

Organizations pursuing large-scale RCM programs typically train their own facilitators who operate in a part-time or full-time role (preferable). These personnel must be carefully selected and trained.

### *Predictive Technology Groups, Lube Groups*

The RCM analysis process will deliver the optimal maintenance program. Part of the task specifications will specify tasks which use predictive technology (e.g. vibration analysis, infra-red thermography, ultrasound, oil analysis etc) where appropriate and the proper specification of lubrication and greasing items. Both of these are often best handled by a specialist group. The organization, change management/training process must be managed to keep up with the program.

## Site Visits, Benchmarking, Conference Attendance

Utilities considering or already involved with should make site visits to other utilities and asset-intensive who have been successful with reliability and excellence programs and attend conferences such as those run by the SMRP to keep abreast of what is happening in the maintenance and reliability world. Benchmarking performance against other utilities is always an interesting exercise.

# APPENDIX A
# LITERATURE SURVEY

## BOOKS

Anderson, J.G. and Edgar, E.C. 1989. *Demonstration of Reliability-Centered Maintenance*. Palo Alto, Calif.: Electric Power Research Institute.

Anderson, R.T. and Lewis, N. 1990. *Reliability-Centered Maintenance: Management and Engineering Methods*. London and New York: Elsevier Applied Science.

Anderson, R.T. and Neri, L. 1990. *Reliability-Centered Maintenance: Management and Engineering Methods*. London; New York: Elsevier Applied Science; New York, NY: Elsevier Science Pub. Co.

Andrews, J.D. and Moss, T.R. 1993. *Reliability and Risk Assessment*. Harlow, Essex: Longman.

August, J. 2004. *RCM Guidebook: Building a Reliable Plant Maintenance Program*. Tulsa, Okla.: PennWell.

August, J. 1999. *Applied Reliability Centered Maintenance*. Tulsa, Okla.: PennWell.

Berry, J.E. 1996. *Detection of Multiple Cracked Rotor Bars on Induction Motors Using Both Vibration and Motor Current Analysis*. P/PM Technology.

Bigliardi, Louis A. 1990. *Glossary of Refining Terms*. Detroit, MI.

Blanchard, B.S. 1986. *Logistics Engineering and Management*. Englewood Cliffs, NJ.: Prentice-Hall.

Blanchard, B.S. and Fabrycky, W.J. 1990. *Systems Engineering and Management*. Englewood Cliffs, NJ: Prentice-Hall.

Blanchard, B.S., Verma, D. and Peterson, E.L. 1995. *Maintainability: A Key to Effective Serviceability and Maintenance Management*. New York, NY.: John Wiley and Sons.

Blanchard, B.S., Verma, D. and Peterson, E.L. 1995. *Maintainability*. New York, NY: Wiley.

Bloom Neil B. 2005. *Reliability Centered Maintenance – Implementation Made Simple*. New York, NY.: McGraw-Hill.

Camp, Robert, C. 1989. *Benchmarking: The Search for Industry Best Practices that Lead to Superior Performance*. Milwaukee, WI.: ASQC Quality Press.

Cartin, Thomas J. 1993. *Principles and Practices of TQM*. Milwaukee, WI.: SQC Quality Press.

Chalifoux, A. and Baird, J. 1999-04. *Reliability Centered Maintenance (RCM) Guide; Operating a More Effective Maintenance Program*. Champaign, Ill.: US Army Corps of Engineers, Construction Engineering Research Laboratories.

Chalifoux, A. and Baird, J. 1999. *Reliability Centered Maintenance (RCM) Guide; Operating a More Effective Maintenance Program*. US Army Corps of Engineers, Construction Engineering Research Laboratory; Springfield, VA: National Technical Information Service.

Chalifoux, A., Northrup, J.I. and Chan, N.Y. 1999. *Streamlined Reliability Centered Maintenance (RCM)*. US Army Corps of Engineers, Construction Engineering Research Laboratory; Springfield, VA: National Technical Information Service.

Cox, S.J. and Tait, N.R.S. 1991. *Reliability, Safety and Risk Management*. Oxford: Butterworth Heinemann.

Davis, D.J. 1952. *An Analysis of Some Failure Data*. Journal of the American Statistical Association.

Del Zingaro, M. and Matthews, C. 1995. *Using Power Signature Analysis to Detect the Behavior of Electric Motors and Motor Driven Machines*. P/PM Technology.

Department of Defense. 1984. *Procedures for Performing a Failure Mode, Effects and Critical Analysis*. Washington DC, Military Standard MIL-DTD, 1629A, Notice 2.

Dept of Defense. 1980. *Guide to reliability centered maintenance (RCM) for fielded equipment*. [Washington: Dept. of Defense], Dept. of the Army.

Dovich, Robert and Wortmann, Bill. 1995. *Certified Reliability Engineer Primer*. Quality Council of Indiana.

Electric Utility Consultants. 2000. *Reliability Centered Maintenance (RCM) for Substation, Transmission and Distribution* Conference, May 25-26, 2000, Westin Hotel, Denver, Colorado. Calgary: Power System Solutions.

Electric Utility Consultants. 2000. *Reliability Centered Maintenance (RCM) in a Competitive Business Environment* Conference, May 23-24, 2000, Westin Hotel, Denver, Colorado. Calgary: Power System Solutions.

Fitzhenry, Robert I. 1993. *The Harper Book of Quotations*. Harper Collins.

Flanagan, D.M. 1991. *Software Considerations for Reliability Centered Maintenance Analyses*. Oak Ridge, Tenn.: Oak Ridge National Laboratory.

Fogarty, Donald W., Hoffmann, Thomas R. and Stonebaker, Peter. 1989. *Production and Operations Management*. Cincinnati, OH: South-Western Publishing Co.

Gaertner, J.P. 1989. *Demonstration of Reliability Centered Maintenance*. Palo Alto, CA: Electric Power Research.

Gano, Dean L. 1993. *Incident Investigation and Problem Solving Techniques*. Richland, WA: Appollonian Publications.

Gleick, J. 1987. *Chaos-Making a New Science*. New York: Penguin.

Goble, W.M. 2003. *Accurate Failure Metrics for Mechanical Instruments*. NC: Research Triangle Park: ISA.

Goble, W.M. 1998. *Control Systems Safety and Evaluation and Reliability*. NC: Research Triangle Park: ISA.

Hammer, Micheal and Champy James. 1993. *Re-Engineering the Corporation*. New York: Harper Collins.

Hänninen, S. and Laakso, K. 1993. *Experience-based Reliability Centered Maintenance: An Application on Motor Operated Valve Drives*. Helsinki: Finnish Centre for Radiation and Nuclear Safety.

Hemmelman, K.M. 1992. *Reliability Centered Maintenance of Power Transformers and Circuit Breakers*. Oak Ridge, Tenn.: Oak Ridge National Laboratory.

Hill, E.S., Midgett, W.D. and Schwan, C.A. 1997. *Reliability Centered Maintenance (RCM) Technical Reference for Power Delivery*. Palo Alto, CA: Electric Power Research Institute.

Hutchins, Greg. 1993. *A Comprehensive Guide to Registration, Audit Guidelines and Successful Certification*. Essex Junction, VT: Oliver Wright Publications, Inc.

Irenson, Grant William. 1998. *Handbook of Reliability Engineering and Management*. New York.: McGraw-Hill.

Jones, Richard B. 1995. *Risk Based Management: A Reliability Centered Approach*. Houston, TX: Gulf Publishing.

Krishnamoorti, K.S. 1992. *Reliability Methods for Engineers*. Milwaukee, WI: ASQC Quality Press.

Lammintausta, M. and Lehtonen, M. 2003. *Reliability Centered Maintenance (RCM): Application to Circuit Breakers*. Espoo: Helsinki University of Technology.

Levitt, J. 2005. *Managing Factory Maintenance*. New York: Industrial Press.

Maintenance Steering Group-3 Task Force. 1993. *Maintenance Program Development Document*. Washington DC: Air Transport Association of America.

Midgett, W.D. 1991. *Guide for Monitoring Effectiveness of Utility Reliability Centered Maintenance (RCM) Programs; Final Report*. Palo Alto, Calif.: The Institute.

Mohr, G. 1995. *Technology Overview-Ultrasonic Detection*. P/PM Technology.

N/A. 1995. Dependability Management-Part 3-11: *Application Guide-Reliability Centered Maintenance*. Geneva, Document no. 56/651/FDIS.

Nakajima, Seiichi. 1988. *Introduction to TPM*. Cambridge, MA: Productivity Press.

Nakajima, Seiichi. 1988. *TPM Development Program*. Cambridge, MA: Productivity Press.

Nelson, W. 1982. *Applied Life Data Analysis*. New York: Wiley.

Niaki, S. and Broscious, J.A. 1985. *Underground Tank Leak Detection Methods-A State of the Art Review*. Springfield Virginia: National Technical Information Service, US Department of Commerce.

Nowlan, F.S, and Heap, H. 1978. *Reliability Centered Maintenance*. Springfield Virginia: National Technical Information Service, US Department of Commerce.

Oakland, J.S. 1984. *Total Quality Management*. New York: Harper Collins.

Perrow, C. 1984. *Normal Accidents*. New York: Harper Collins.

Purucker, S.L. 1992. *Substation Reliability Centered Maintenance (RCM)*. Oak Ridge, Tenn.: Oak Ridge National Laboratory.

Reason, J.T. 1990. *Human Error*. Cambridge England: Cambridge University Press.

Resnikoff, H.L. 1978. *Mathematical Aspects of Reliability Centered Maintenance*. Los Altos, CA: Dolby Access Press.

Resnikoff, H.L. 1979. *Mathematical Aspects of Reliability-Centered Maintenance*. Los Altos, Calif.: Dolby Access Press; Springfield, Va.: reproduced by National Technical Information Service.

Robinson, Charles and Ginder, Andrew Paul. 1995. *Implementing TPM: The North American Experience*. Portland, OR: Productivity Press.

Sandtorv, H. and Rausand, M. 1991. *Closing the Loop between Design Reliability and Operational Reliability*. Maintenance, 6(1), 13-21.

Schwan C. 1996. *Reliability Centered Maintenance (RCM) Technical Reference for Substations*. Palo Alto, CA: Electric Power Research Institute.

Schwan, C.A. 1991. *Guide for Generic Application of Reliability Centered Maintenance (RCM) Recommendations*. Palo Alto, Calif.: The Institute.

Schwan, C.A., Morgan, T.A., Edgar, E.C. and Betros, J.R. 1991. *Demonstration of Reliability Centered Maintenance: Final Report*. Palo Alto, Calif.: Electric Power Research Institute.

Smith, A.M. 1993a. *Reliability-Centered Maintenance*. New York: McGraw-Hill.

Smith A.M., Hinchcliffe G.R., 2004b *RCM – Gateway to World Class Maintenance*. Burlington, MA.: Elsevier Butterworth-Heinemann

Smith, D.J. 2005. *Reliability, Maintainability and Risk: Practical Methods for Engineers Including Reliability Centred Maintenance and Safety-Related Systems*. Oxford: Butterworth-Heinemann.

Smith, D.J. 1993. *Reliability, Maintainability and Risk*. Oxford: Butterworth Heinemann.

Snow, D.A. 1991. *Plant Engineer's Reference Book*. Oxford: Butterworth Heinemann.

Suzuki, Tokutaro. 1994. *TPM in Process Industries*. Portland, OR: Productivity Press.

Tatikonda, Lakshmi U. and Tatikonda, Rao J. 1996. *Measuring and Reporting the Cost of Quality*, Production and Inventory Management Journal, Second Quarter 1996 Vol 37, No. 2 to 7.

Toms, L.A. 1995. *Machinery Oil Analysis*. Pensacola Florida: Published by Author.

Tsang, A.H.C. 2000. *Reliability Centred Maintenance: A Key to Maintenance Excellence = Yi ke kao xing wei zhong xin de wei hu bao yang: mai xiang zhuo yue zhi dao*. Hong Kong: City University of Hong Kong.

Weaver, C. 1995. *Time Waveform Analysis*. P/PM Technology.

Westerkamp, Thomas A. 1992. *Plant Manager's Standard Manual and Guide*. Englewood Cliffs, NJ: Prentice-Hall.

Wireman, T. 1998. *Developing Performance Indicators for Managing Maintenance*. Industrial Press Ind, NY.

Wong, D. 2001. A *Knowledge-Based Decision Support System in Reliability-Centered Maintenance of HVAC Systems*. Ottawa: National Library of Canada = Biblioth_que nationale du Canada.

Xu, Ming and Blue, J. 1995. *Condition Monitoring of Seal-less Pumps Using Spike Energy*. P/PM Technology.

Zwingelstein, G. 1996. *Reliability Centered Maintenance, A Practical Guide for Implementation*. Paris: Hermes.

## CONFERENCES

Baker, T.T. 2004. Modern On Line Electric Motor Monitoring-A New Realm in Predictive Maintenance. SMRP Conference, Norfolk, Virginia.

Bass, K. 2004. Addressing the Soft Side Results in Hard Dollars. SMRP Conference, Norfolk, Virginia.

Beck, B. and Clark, D. 2003. Incorporating Maintenance Information into Capital Projects. SMRP Conference, Indianapolis, Indiana.

Beebe, R. 2003. Condition Monitoring By Performance Analysis. SMRP Conference, Indianapolis, Indiana.

Bergman, W.J. 2000. Reliability Centered Maintenance 101 Workshop: May 22, 2000, Westin Hotel, Denver, Colorado. Calgary: Power System Solutions.

Bowers, S.V. 1995. Integrated Strategy for Predictive Maintenance of AC Induction Motors. Predictive Maintenance Technology National Conference, Indianapolis, Indiana.

Branham, D.C. 2004. Oil Mist Lubrication-A Must for Reliability Improvement. SMRP Conference, Norfolk, Virginia.

Carson, S. 2003. Internet Based Machinery Analysis. SMRP Conference Indianapolis, Indiana.

Cline, B. 2004. Stop the Insanity. SMRP Conference, Norfolk, Virginia.

Dabbs, T. 2003. How to Make a Transition from Reactive to Proactive Maintenance and Financial Benefits Associated with the Improvements. SMRP Conference, Indianapolis, Indiana.

Dabbs, T. 2004. Three Waves to Reliability: Process to Establish Reliability. SMRP Conference, Norfolk, Virginia.

Dalley, R. 1995. Wear Particle Analysis. Predictive Maintenance Technology National Conference, Indianapolis, Indiana.

Davidson, T. 2004. Essentaial Components of Effective Maintenance Planning. SMRP Conference, Norfolk, Virginia.

Davis, J. 2003. Creating a Reliability Culture in Your Company. SMRP Conference, Indianapolis, Indiana.

Deane, T. and Baker, P. 2003. Using Data Analysis to Identify Business Improvement Opportunities. SMRP Conference Indianapolis, Indiana.

Ducci, PP. 2003. The Hidden Benefits From Taking a Process View of Asset Management. SMRP Conference, Indianapolis, Indiana.

Ebersole, E and Newhard, E. 2004. A Systems Approach to Failure Modes and Effects Analysis. SMRP Conference, Norfolk, Virginia.

Eisenbise and Schrader. 2003. Root Cause Failure Analysis-A Key Element of Systematic Failure Elimination. SMRP Conference, Indianapolis, Indiana.

Fynn, C. 2005. Benefits of Utilizing SAP EAM and Reliability (RCM) Best Practice in Design-Build Phases of Capital Project/Asset Life-Cycle. Houston, Texas; Tucson, AZ: SAP-Centric EAM 2005.

Goldstein, M.R. 2004. Maintenance Productivity, Correcting What the Critics Say. SMRP Conference, Norfolk, Virginia.

Granat, M. 2004. Roadmap to the Right Hire, How to Hire and Keep the Star Performers. SMRP Conference, Norfolk, Virginia.

Gulati, R. 2004. Creating and Sustaining Reliability Culture. SMRP Conference, Norfolk, Virginia.

Gwaltney, R. 2003. The Systematical Approach to Technical Training. SMRP Conference, Indianapolis, Indiana.

Hanschom, M. 2004. Data Collection for Maintenance Management by Information. SMRP Conference, Norfolk, Virginia.

Haverley, J. 2004. Use of Advanced Ulntrasonic Techniques for Weld Inspections and Life Extension. SMRP Conference, Norfolk, Virginia.

Henson, G.E. 2003. Benefits and Lessons Learned from Predictive Technologies at Clinton Labs. SMRP Conference, Indianapolis, Indiana.

Himmesoete, D. and Britton, K. 2003. Performance Contracting. SMRP Conference, Indianapolis, Indiana.

Huysman, W. 2004. Utilizing Analytical Ferrography for Root Cause Analysis and Failure Prevention. SMRP Conference, Norfolk, Virginia.

Idhammar, C. 2004. Results Orientated Reliability and Maintenance Management. SMRP Conference, Norfolk, Virginia.

Isenhour, S. 2004. Reliability is Not Just About the Machines, Developing Your Other Assets....People. SMRP Conference, Norfolk, Virginia.

James, R. 1995. Basic Oil Analysis. Predictive Maintenance Technology National Conference, Indianapolis, Indiana.

Kalinouskas, R. and McDonald, C.A. 2004. Developing and Effective Root Cause Analysis Work Process. SMRP Conference, Norfolk, Virginia.

Kane, C.F. 1996. Predictive Maintenance Technologies Can Help Prioritize Maintenance Dollars. Northwest Indiana Business Roundtable and Trade Show, Merriville, Indiana, 3 October 1996.

Latino, K. 2003. Utilizing Advanced Statistical Reliability Methods to Improve Overall Asset Performance. SMRP Conference, Indianapolis, Indiana.

Leonard, R. 2004. New Generation High Speed Camera Technology. SMRP Conference, Norfolk, Virginia.

MacGuire, B. 2004. Work Smart to Achieve a Common Asset Reliability Strategy. SMRP Conference, Norfolk, Virginia.
Mariarty, T.J. 2004. The Importance Of Process Discipline. SMRP Conference, Norfolk, Virginia.
McElroy, J.W. 2003. Reliability Risk-Based Maintenance. SMRP Conference, Indianapolis, Indiana.
McFarland, R.P. 2003. Metrics for Managing Maintenance. SMRP Conference, Indianapolis, Indiana.
Moubray, J.M. 1987. Reliability Centered Maintenance. A Conference on Condition Monitoring, Gol, Norway.
Moubray, J.M. 1989. Maintenance and Safety-A Pro-Active Approach. Annual Conference of the Accident Prevention and Advisory Unit of the United Kingdom National Health and Safety Executive, Liverpool, UK.
Moubray, J.M. 1989. Maintenance and Product Quality. International Conference on Total Quality, Hong Kong.
Moubray, J.M. 1988. Maintenance Management-The Third Generation. The 9th European Maintenance Congress, Helsinki, Finland.
Moubray, J.M. 1988. Developments of Reliability Centered Maintenance. The Factory Efficiency & Maintenance Show and Conference, NEC, Birmingham, UK.
Moubray, J.M. 1998. The Responsible Custodianship of Physical Assets. The Tenth Annual Canadian Maintenance Management Conference, Toronto, Canada.
Moubray, J.M. 1995. Maintenance Management-A New Paradigm. Third Annual Conference of the Society of Maintenance and Reliability Professionals, Chicago, Illinois.
Netherton, D. 2003. Optimizing Maintenance Task Intervals: The Agony and the Ecstasy. SMRP Conference, Indianapolis, Indiana.
Nicholas, J. 2003. Overview of RCM including its Variants and Derivatives. SMRP Conference, Indianapolis, Indiana.
Nicholas, J.R. (jr). 1995. AC and DC Motor Circuit Testing and Predictive Analysis. Predictive Maintenance Technology National Conference, Indianapolis, Indiana.
Nicholas, J.R. jr. 2003. Settling the Controversy about Reliability Centered Maintenance, It's Variants and Derivatives. SMRP Conference, Indianapolis, Indiana.
Nichols, J. and Ellis, K. 2005. RCMO-Integrated Best Practice EAM and Reliability With SAP PM. Houston, Tucson, AZ: SAP-Centric EAM 2005.
O'Hanlon, T. 2003. CMMS Best Practices. SMRP Conference, Indianapolis, Indiana.
Oliver, R. 2004. The Deviant Cost Model-An Alternative Way to Look at Cost. SMRP Conference, Norfolk, Virginia.
Palarchio, G. 2003. Maximizing Asset Reliability. SMRP Conference, Indianapolis, Indiana.
Parrot, M. 2004. Using an Internet Delivered CMMS to Build a Foundation for Improving Equipment Reliability. SMRP Conference, Norfolk, Virginia.
Peerman, M. 2004. Shorter Maintenance Outages Increase Equipment Reliability. SMRP Conference, Norfolk, Virginia.
Probst, R. 2003. Closing the "Know-Do" Gap. SMRP Conference, Indianapolis, Indiana.
Robinson, Dr J.C. and Piety, Dr K.R. 1996. Peak Value (Peak Vue) Analysis: Advantages Over Demodulation for Gearing Systems and Slow-speed Bearings. Northwest Indiana Business Roundtable and Trade Show, Merriville, Indiana, 3 October 1996.
Rose, A. 1994. How to Set Up an Electrical Predictive Maintenance Program. Predictive Maintenance Technology National Conference, Indianapolis, Indiana.

Sachs, N. 2004. V-Belts, Understanding their Design, Operation and PDM Inspections. SMRP Conference, Norfolk, Virginia.

Smith, M. 2003. A Practical Discussion of RCM Implementation. SMRP Conference, Indianapolis, Indiana.

Snell, L. 2004. Integrating Operations Into Reliability and Maintenance Work Processes. SMRP Conference, Norfolk, Virginia.

Staat, D. and Walker, T. 2003. Improving Maintenance Work Management Processes. SMRP Conference, Indianapolis, Indiana.

Sullivan, R. 2004. Precision Alignment Implementation. SMRP Conference, Norfolk, Virginia.

Toth, L. 2004. RCM at Sabrante Water Treatment Plant: What Did We Find Out and Where Do We Go from Here?. SMRP Conference, Norfolk, Virginia.

Troyer, D. 2003. Manager's Strategy for Increasing Profit Through Lubrication Excellence. SMRP Conference, Indianapolis, Indiana.

Van der Horn, G. and Woyshner, W. 1995. Electric Motor Predictive Maintenance-A Comprehensive Approach. Predictive Maintenance Technology National Conference, Indianapolis, Indiana.

White, G. 1995. Vibration Data Collectors and Analyzers. Predictive Maintenance Technology National Conference, Indianapolis, Indiana.

White, G. 1995. Designing and Installing a Monitoring System. Predictive Maintenance Technology National Conference, Indianapolis, Indiana.

Wilson, Paul F., Dell Larry D and Anderson, Gaylord F. 1993. Root Cause Analysis: A Tool for Total Quality Management. Milwaukee, WI: ASQC Quality Press.

## WEB SITES

ADEPA-UPM. 2000. Study of Existing Reliability Centered Maintenance Approaches Used In Different Industries. <http://www.laas.fr/~kader/rcm2.pdf> [Cited February 9, 2005].

Alexander, D. 2003. Application of Monte Carlo Simulations to System Reliability Analysis. <http://turbolab.tamu.edu/pubs/Pump20/p20pg091.PDF> [Cited January 5, 2005].

Barringer, P. 2002. Predict Future Failures. <http://www.plant-maintenance.com/maintenance_articles_rcm.shtml> [Cited July 18, 2005].

Beckner, W.D. 2002. Metalclad Switchgear Failure and Consequent Losses of Off Site Power. <http://www.nrc.gov/reading-rm/doc-collections/gen-comm/reg-issues/2002/in02001.html> [Cited July 18, 2005].

Bethel, N. 2002. VFD Motor Testing. <http://www.pdma.com/VFDtest.html> [Cited July 18, 2005].

Bever, K. Integration of Plant Enterprise Systems Enables Significant Asset Optimization. <http://domino.automation.rockwell.com/applications/gs/region/EntekWebST.nsf/files/Bever99.pdf/$file/Bever99.pdf> [Cited January 5, 2005].

Bever, K. Integration - Key To Asset Optimization. <http://www.mt-online.com/articles/09-99pm.cfm> [Cited April 6, 2005].

Bloom, N. 2004. Understanding Hidden Failures in the RCM Analysis. <http://www.plant-maintenance.com/maintenance_articles_rcm.shtml> [Cited July 18, 2005].

Canada Correctional Services. 2003. Management of Waste Water Treatment Systems. <http://www.csc-scc.gc.ca/text/plcy/doc/318-6-gl_e.pdf> [Cited January 5, 2005].

Capital Regional District. Unknown. CRD-Supply and Demand Management Executive Summary. <http://www.crd.bc.ca/water/reports/strategic/vol1.htm> [Cited January 5, 2005].

Chatterjee, S.K. 2003. Seal Failures Can Happen Even on Idle Standby Pumps. <http://www.pump-magazine.com/pump_magazine/field_cases/case3/case3.htm> [Cited July 18, 2005].

Chatterjee, S.K. 2003. The Essence of Equipment Failure Analysis-Theory, Approach and Case Study. <http://www.pump-magazine.com/pump_magazine/pump_articles/article_09/article_09.htm> [Cited July 18, 2005].

Chopp, B.M. 2003. Reliability Centered Maintenance. <http://www.energy2003.ee.doe.gov/presentations/om/1-Chopp2.pdf> [Cited January 5, 2005].

Dunn, S. 1997. Best Practice Maintenance Strategies for Mobile Equipment. <http://www.plant-maintenance.com/maintenance_articles_rcm.shtml> [Cited July 18, 2005].

Dunn, S. 1999. Integrating RCM with Effective Planning and Scheduling. <http://www.plant-maintenance.com/maintenance_articles_rcm.shtml> [Cited July 18, 2005].

Dunn, S. Unknown. Maintenance Task Selection. <http://www.plant-maintenance.com/maintenance_articles_rcm.shtml> [Cited July 18, 2005].

Duran, J. 2005. The Role of the Facilitator in Operational Reliability. <http://www.plant-maintenance.com/maintenance_articles_rcm.shtml> [Cited July 18, 2005].

Duran, J. 1998. What is Operational Reliability. <http://www.plant-maintenance.com/maintenance_articles_rcm.shtml> [Cited July 18, 2005].

Electrical Power Research Institute. . 37 Substations. <http://www.epriweb.com/ public/2006_P037.pdf> [Cited January 5, 2005].

Farrel, A. and Burris, J. 1995. Alternatives to Chlorine-Is Ultra Violet the Answer?. <http://www.fwrj.com/articles/9506.pdf> [Cited February 9, 2005].

Fortin, J., Colbert, J. and Regan, T. . Asset Replacement Strategies at the MWRA. <http://www.newea.org/AMRC/6_ass/Fortin,%20Colbert,%20Regan%20Chapter%206%20Asset%20Replacement%20Strategies%20at%20the%20MWRA%20rev%207.doc> [Cited July 18, 2005].

Fortin, J., Vetere, J., O'Brein, D., Gallinaro, G., Regan, T., Colbert, J. . MWRA's Facility Asset Management Program Phase 1. <http://www.mt-online.com/articles/0902_mwraphase1.cfm> [Cited April 6, 2005].

Fortin, J.W., Colbert, J.P., Regan, T. Unknown. Outside Resources Contribute to Culture Change. <http://www.mt-online.com/articles/0902_culturechange.cfm> [Cited April 6, 2005].

Gritzuk, P.E. 2003. The Importance of Water Re-Use in the 21st Century. <http://resourcescommittee.house.gov/108cong/water/2003mar27/gritzuk.htm> [Cited January 5, 2005].

Hazardous Installation Directorate. 2001. Reliability of Utilities. <http://www.hse.gov.uk/comah/sragtech/techmeasutilitie.htm> [Cited February 9, 2005].

Henry, J.L. 1996. High Level of Reliability Relies on a High Level of Maintenance. <http://tdworld.com/mag/power_high_level_reliability/> [Cited January 5, 2005].

Higgs, P.A., Parkin, R. and Jackson, M. 2004. A Survey on Condition Monitoring Systems in the Industry. <http://www.plant-maintenance.com/maintenance_articles_rcm.shtml> [Cited July 18, 2005].

Howell, A., Hanson, K., Dhole, V. and Sim, E. 2002. Optimizing Asset Utilization through Process Engineering. <http://www.cepmagazine.org/pdf/090254.pdf> [Cited January 5, 2005].

Hughes, R.L. 2002. Analysis: RCM vs RCA. <http://www.plant-maintenance.com/maintenance_articles_rcm.shtml> [Cited July 18, 2005].

Idhammer, T. Unknown. Checking Best Practices for PM. <http://www.plant-maintenance.com/maintenance_articles_rcm.shtml> [Cited July 18, 2005].

Jakovljevic P.J. and Thompson O. 2004. Reliability Drive Maintenance. <http://www.cmmscity.com/articles/reliability_driven_maintenance.htm> [Cited January 5, 2005].

Jegnan, J. 2005. When the Power Must Stay On. <http://www.facilitiesnet.com/ ms/article.asp?id=3000> [Cited February 9, 2005].

Kasper, R.J. 1980. Providing Reliable Switchgear Control Power. <http://www.sandc.com/edocs_pdfs/EDOC_001715.pdf> [Cited July 18, 2005].

Keeter, B. Unknown. Managing Availability for Improved Bottom Line Results. <http://www.plant-maintenance.com/maintenance_articles_rcm.shtml> [Cited July 18, 2005].

Kindrachuk, G. . Integration of asset management and GIS. <http://www.gisdevelopment.net/proceedings/gita/2001/tat/techgi088pf.htm> [Cited February 9, 2005].

Latimo, M. 1999. RCFA+RCM=Formula for Successful Maintenance. <http://www.plant-maintenance.com/maintenance_articles_rcm.shtml> [Cited July 18, 2005].

Latimo, M. Unknown. The Failure Dilemma. <http://www.plant-maintenance.com/maintenance_articles_rcm.shtml> [Cited July 18, 2005].

Latino, C.J. Unknown. Hidden Treasure: Eliminating Chronic Failures Can Cut Maintenance Costs Up to 60%. <http://www.plant-maintenance.com/maintenance_articles_rcm.shtml> [Cited July 18, 2005].

Monitor Europe. Unknown. RCM-Glossary and Abbreviations. <http://www.monitoreurope.com/glossary/> [Cited February 9, 2005].

Moubray, J.M. 2005. 21st Century Maintenance Organization. <http://www.mt-online.com/articles/moubray_21stcentury.cfm> [Cited July 18, 2005].

MRO Software. 2003. Strategic Asset Management in the Utility Industry. <http://www.mro.com/corporate/pdf/Utilities_whitepaper_final.pdf> [Cited April 6, 2005].

MTE Corp. Unknown. Solving Motor Failures Due to Peak Voltages and Fast Rise Times. <http://www.mtecorp.com/solving.html> [Cited April 6, 2005].

Nailen, R. 2005. Is a Repaired Motor Less Reliable. <http://www.maintenanceworld.com/Articles/NailenR/repaired-motor-less-reliable.htm> [Cited February 9, 2005].

NASA. 2001. Standardized Facilities Preventative Maintenance Work Task Guide. <http://www.hq.nasa.gov/office/codej/codejx/PMMaintWorkTaskGuide.doc> [Cited February 9, 2005].

NEA/CSNI. 1999. Common Cause Failure of Centrifugal Pumps. <http://www.nea.fr/ html/nsd/docs/1999/csni-r99-2.pdf> [Cited July 18, 2005].

NEA/CSNI. 2003. Collection of Common Cause Failure of Check Valves. <http://www.nea.fr/html/nsd/docs/2003/csni-r2003-15.pdf> [Cited July 18, 2005].

Nicholas, J.R., Jr., P.E., CMRP CEO MQS LLC., 2004. Keynote Address To the International Maintenance Conference, Bonita Springs, Florida, USA , December 6, 2004, <http://www.reliabilityweb.com/art04/back_to_basics.htm> [Cited July 18, 2005].

Nichols, J. and Mathusheski, B. 2000. Maintaining Substation Reliability. <http://www.plant-maintenance.com/maintenance_articles_rcm.shtml> [Cited July 18, 2005].

Nilsen. Unknown. Your Switchgear is Your Protection, Right? Is It?. <http://www.nilsen.com.au/nilsen/rwpattach.nsf/viewasattachmentPersonal/3C9D9C2D2ABCBFE5CA256B4100797BF0/$file/r98I3V2-page%2011.pdf> [Cited July 18, 2005].

NWRC and MS (Canada). 2001. Threats to Water Availability in Canada. <http://www.nwri.ca/threats2full/ch5-1-e.html> [Cited July 18, 2005].

Palmer, J. 2004. Switchgear Malfunction Results in Electrician's Death. <http://www.ecmweb.com/mag/electric_switchgear_malfunction_results/> [Cited January 5, 2005].

Penrose, H. Unknown. Basic Overview of RCM Based Approach for Motor Management Programs. <http://www.plant-maintenance.com/maintenance_articles_rcm.shtml> [Cited July 18, 2005].

Plant Maintenance Resource Center. 2005. Equipment Maintenance Strategies and RCM. <http://www.plant-maintenance.com/maintenance_articles_rcm.shtml> [Cited July 18, 2005].

Plucknette, DJ. 2004. When and How to Template a RCM Analysis. <http://www.reliabilityweb.com/art04/rcm_template.htm> [Cited July 18, 2005].

Polka, D. 2001. How to Maintain a Variable Frequency Drive. <http://cementamericas.com/mag/cement_maintain_variable_frequency/> [Cited January 5, 2005].

Pomorski, T.R. 2004. Total Productive Maintenance (TPM) Concepts and Literature Review. <http://www.brooks.com/documents.cfm?documentID=2110> [Cited January 5, 2005].

Post, R.L, Hendershot, D.C. and Kers, P. 2001. Safety and Reliability-A Synergetic Approach. <http://home.att.net/~d.c.hendershot/papers/pdfs/aiches01ppss.pdf> [Cited January 5, 2005].

Post, R.L., Hendershot, D.C. and Kers, P. 2002. Synergetic Design-Approach to Safety and Reliability Yields Great Benefits. <http://www.cepmagazine.org/pdf/010260.pdf> [Cited January 5, 2005].

Rahkonen, O.H. and Mangs, J. 2002. Fire Induced Damage to Electrical Cables and Fire Growth on Cables. <http://www.eurosafe-forum.org/ipsn/pdf/euro2_1_4_fireinduced_damage.pdf> [Cited January 5, 2005].

Rand Water Board. 2001. Rand Water Board Chief Executive's Report. <http://www.randwater.co.za/pdf/Annual2001/ChiefExecutives2001part2.pdf> [Cited July 18, 2005].

Sahdev, M. Unknown. General Symptoms of Cavitation and its Affect on Pump Performance and Pump Parts. <http://www.cheresources.com/centrifugalpumps3b.shtml> [Cited January 5, 2005].

Schimmoller, BK. 1998. Skake Those Boeing Blues. <http://www.oit.doe.gov/bestpractices/energymatters/emextra/boeing.shtml> [Cited July 18, 2005].

Tissue, B.M. 1996. SCIMEDIA-Analytical Chemistry and Instrumentation. <http://www.scimedia.com> [Cited August 04, 2005].

Traianou, C. 1999. Reliability Improvement for the New Millennium. <http://www.reliabilityweb.com/excerpts/excerpts/WATER%20CORPORATION.pdf> [Cited July 18, 2005].

Troyer, D. 2004. Establishing Moisture Contamination Targets for Hydraulic Systems. <http://www.machinerylubrication.com/article_detail.asp?articleid=573&relatedbookgroup=Lubrication> [Cited February 9, 2005].

Troyer, D. Unknown. How to Lube Up Your FMEA Process. <http://www.plant-maintenance.com/maintenance_articles_rcm.shtml> [Cited July 18, 2005].

Troyer, D. Unknown. RCM and it's Meaning to the Oil Analysis Professional. <http://www.plant-maintenance.com/maintenance_articles_rcm.shtml> [Cited July 18, 2005].

Turkel, S.S. 1995. Understanding Variable Speed Drives. <http://www.ecmweb.com/mag/electric_understanding_variable_speed_2/> [Cited January 5, 2005].

UE Systems. 2002. Technology Overview: Ultrasonic Detection. <http://uesystems.com/techview.htm> [Cited January 5, 2005].

Unknown. 2002. Component Vulnerability Modeling. <http://www.crid.or.cr/crid/CD_Agua/pdf/eng/doc14607/doc14607-4.pdf> [Cited January 5, 2005].

Unknown. 1998. Biological Treatment. <http://www.ecy.wa.gov/pubs/9837/t3.pdf> [Cited January 5, 2005].

Unknown. 2004. News Letter. <http://www.ema-inc.com/communicators/com_spring04.htm> [Cited January 5, 2005].

Unknown. 2004. Selecting Instrumentation Equipment for a Safety Application. <http://www.exida.com/articles/AssessmentLevels.pdf> [Cited February 9, 2005].

Unknown. 2000. What is Pro-Active Maintenance. <http://www.plant-maintenance.com/maintenance_articles_rcm.shtml> [Cited July 18, 2005].

Vahlstrom, W. 2002. Maintaining Your Switch Gear. <http://www.ecmweb.com/mag/electric_maintaining_switchgear_2/> [Cited January 5, 2005].

Watkins, J.P. 1999. RCM-Where does it fit?. <http://domino.automation.rockwell.com/applications/gs/region/EntekWebST.nsf/files/Watkins99.pdf/$file/Watkins99.pdf> [Cited January 5, 2005].

Wiseman, M. 2004. Can RCM and Streamlined RCM Peacefully Co-exist. <http://www.reliabilityweb.com/art04/rcm_streamlined.htm> [Cited July 18, 2005].

## THESIS/DISSERTATION/MANUSCRIPTS

Beyerle, J.F. 1998. Reliability Centered Maintenance for 12KV Switchgear Control, Meter and Protective Relay Systems. Thesis/dissertation/manuscript Archival Material viii, 141 leaves. University of Louisville Database: WorldCat

Knottek, J.A. 1998. Evaluation of Reliability-Centered Maintenance Techniques for Electric Substation Applications. Thesis/dissertation/manuscript Archival Material ix, 108 p. University of Missouri–Rolla Database: WorldCat

Rowley, P.J. 2003. The Application of Reliability Centered Maintenance to the Water Treatment Environment. Thesis/dissertation/manuscript Archival Material vii, 78 leaves. California State University, Northridge Database: WorldCat

Coetzee, Casper L. 2002. An Optimized Instrument for Designing a Maintenance Plan: A Sequel to Reliability Centred Maintenance. Thesis/dissertation/manuscript. University of Pretoria (South Africa)

Wong, Daniel K Y. 2000. A Knowledge-Based Decision Support System in Reliability-Centered Maintenance of HVAC Systems. Thesis/dissertation/manuscript. Memorial University of Newfoundland (Canada)

Tetsu, L. S., 1995. Understanding Variable Speed Drives <http://www.ecmweb.com/mag/electric_understanding_variable_speed_2> [Cited January 5, 2005].

[illegible] Systems, 2002. Technology Overview: Ultrasonic Detection <http://uesystems.com/ [illegible]> [Cited January 5, 2005].

Unknown, 2002. Component Vulnerability Modeling <http://www.crid.or.cr/crid/CD_Agua/pdf/eng/doc11607/doc11607-1.pdf> [Cited January 5, 2005].

Unknown, 1995. Biological Treatment <http://www.epa.gov/[illegible].pdf> [Cited January 5, 2005].

Unknown, 2004. News Letter <http://www.[illegible].com/[illegible]_spring04.htm> [Cited January 5, 2005].

Unknown, 2004. Selecting Instrumentation Equipment for a Safety Application <http://www.[illegible].com/articles/AssessmentLevels.pdf> [Cited February 9, 2005].

Unknown, 2000. What is Pro-Active Maintenance <http://www.plant-maintenance.com/maintenance_articles_[illegible].shtml> [Cited July 18, 2005].

Vahlstrom, W., 2002. Maintaining Your Switch Gear <http://www.ecmweb.com/mag/electric_maintaining_switchgear_2/> [Cited January 5, 2005].

Watkins, J.R., 1999. RCM Where does it fit? <http://domino.automation.rockwell.com/applications/[illegible]/Watkins99.pdf/$file/Watkins99.pdf> [Cited January 5, 2005].

Wiseman, M., 2003. Classic RCM and Streamlined RCM: Peacefully Co-exist? <http://[illegible]/streamlined.htm> [Cited July 8, 2005].

## UNPUBLISHED DISSERTATION MANUSCRIPTS

Batalie, [illegible] 2005. Reliability Centered Maintenance for 12KV Switchgear Control, Motor and Protective Relay Systems. Thesis/dissertation manuscript. Archival Material. Louisville, KY: University of Louisville Database: Worldcat.

Kamal, [illegible] A., 1995. Evaluation of Reliability-Centered Maintenance Techniques for Electrical [illegible]. Thesis/dissertation manuscript. Archival Material. [illegible]. Database: WorldCat.

[illegible]

# APPENDIX B
# UTILITY SURVEY ON APPLICABILITY OF RCM TO THE WATER INDUSTRY

Survey Performed by:
New Dimension Solutions
*734 Franklin Avenue, #514*
*Garden City*
*NY 11530*

## ACKNOWLEDGMENTS

The authors of this report are indebted to the following individuals for their cooperation and participation in this project:

- Chris Fynn, New Dimension Solutions
- Steve Sinkoff, New Dimension Solutions
- Marius Basson, New Dimension Solutions
- Peter Mostovac, New Dimension Solutions
- Paul Musca, Decima Research Inc.
- Rick Nadeau, Decima Research Inc.

# EXECUTIVE SUMMARY

A total of 177 respondents representing water utilities across the United States and in Canada completed an online survey that was developed by New Dimension Solutions. The survey instrument was designed to establish personal data of the respondent, develop a profile of the utility that the respondent represents, determine maintenance programs and efforts currently made within their facilities, and ascertain their involvement with and any benefits achieved through the application of Reliability-Centered Maintenance (RCM).

The survey participants included 177 respondents with a representative mix of utility sizes for <10 MGD to >300 MGD. The majority of the respondents were city- or municipality-owned (77%). The majority performed their own maintenance (>85%).

## Current Maintenance / Asset Management Performance and Priority

Data revealed that there is an emergent focus by utility management on asset performance and optimal management of the utility's assets. Over half of respondents (57%) indicated that this focus is becoming an important issue for utility management, and 13% of respondents stated that optimal management and performance of the utility's assets is becoming ***a very critical issue for utility management***. Respondents were also asked to provide reasons for any change in management prioritization of physical asset maintenance and performance. When the first two reasons for this change in management prioritization were grouped together, aging equipment, facilities, or infrastructure (21%), the desire for better performance, higher efficiency, or optimization (16%), the need for cost reduction or to control costs (10%), and for compliance with regulatory requirements or more stringent regulations (10%) were the most commonly mentioned responses.

Respondents were also asked to specify the most serious issues that currently face utility management, in terms of the maintenance of its assets. When the top three most serious issues were grouped together, capital replacement costs (45% included this in their top three) was the most commonly mentioned, followed by maintenance costs (44%), aging facilities (42%), and reliability (40%).

To gain an understanding of the manner in which maintenance departments function across the various utilities, respondents were asked to select the mode of operation of their maintenance department. A majority of respondents (61%) indicated that their maintenance department function operates in a steadily improving proactive mode, recognizing that some asset systems and facilities continue to be a problem. Twenty-two percent (22%) of respondents say that their maintenance department function operates with the basics of a proactive program, but that maintenance is still mostly reactive in nature.

Plant breakdowns are defined in the survey as being unscheduled failures of an asset that affect plant operations and must be rectified immediately when the asset fails. This is an extremely important issue for utilities with highly loaded or critical assets. According to the survey, on average, 20% of maintenance work is in direct response to these plant breakdowns.

Managing assets via a Computer Maintenance Management System (CMMS) or an Enterprise Asset Management (EAM) system involves tagging physical assets and registration of these assets within the system. The degree to which utilities tagged physical assets and registered these assets within the CMMS / EAM system differed across respondents. For example, 35% of respondents said that the tagging and registering of some critical assets have been completed, yet there was still much more work to be done. Thirty-one percent (31%) indicated that the tagging and registering of most assets has been completed, including all critical assets, with some clean-up work still to be done. ***Another 31% said that no assets at all have been tagged or registered.***

Fixed-interval maintenance tasks (also known as preventive maintenance tasks) are developed from several methods including OEM recommendations, equipment history, formalized preventive maintenance optimization processes, and RCM. When asked to specify the percentage that each of these methods have contributed to the development of the fixed-interval tasks at their utility, respondents on average assigned 48% to OEM recommendations, 31% to equipment history, 10% to formalized preventive maintenance optimization process, and 3% to RCM.

**Asset Management Programs – Key Points From Survey**

Asset management is becoming an important issue with 70% of water utilities. The majority of water utilities believe there is room for improvement in their current utility asset management program and that they are steadily working to improve their programs. Senior management commitment to asset management was developing or strong for greater than two thirds of respondents.

The asset management program for the majority of utilities is in the starting or initial phases with another 22% with no asset management program. The majority of critical assets were included in the asset management programs. The utilities considered that the overall asset condition was the same or improving at over 75% of all utilities with 22% stating that asset condition was deteriorating.

The top three reasons for the formation of an asset management program are:

- Aging infrastructure
- Better performance
- Cost-savings/Reductions

**Maintenance Programs / Failure Management Policies**

Formal and systematic processes and methods to derive maintenance programs generally appear to be in place in many facilities. Forty-five percent (45%) of respondents indicated that these methods have been deployed or utilized at their utility on certain critical assets, and 14% said that a comprehensive program utilizing formal processes to derive asset maintenance programs is in place across the utility. Another quarter (26%) stated that there is a sporadic effort to implement formal and systematic processes / methods at their utility, and 15% said that there are no formal or systematic processes / methods in place at the utility to facilitate the development of maintenance programs for the utility's assets.

Respondents were asked to indicate their level of familiarity with the main Maintenance Strategy/Program Formulation Techniques currently available for developing optimized maintenance programs for physical assets. Familiarity with Original Equipment Manufacturer (OEM) Programs was highest (74%) followed by PM Optimization (46%), RCM (36%), and Root Cause Analysis (34%). In addition to these, respondents were asked to specify which maintenance strategy formulation techniques were used to develop maintenance programs for critical and non-critical physical assets. In both of these cases, the Original Equipment Manufacturer Program was used on the highest percentage of physical assets, followed by PM Optimization, and then RCM.

Applying formalized maintenance strategy formulation techniques or reliability optimization methodologies during the design-build phase of new assets can optimize future maintenance of assets and minimize life-cycle costs. Respondents were asked whether such Maintenance Strategy/program Formulation or Reliability Optimization techniques are applied at their utilities either during the design-build phase of new assets, or during the modification of existing assets. Approximately 44% of respondents indicated that the maintenance staff performs some formalized reliability review of the proposed design, and 31% indicated that design engineers and draughtsman try to consider reliability and maintainability during design. About 22% stated that no formalized maintenance strategy formulation techniques or reliability optimization methodologies are applied in the design-build phase or when modifying existing assets.

With respect to formalized Criticality Analyses for the technical systems or assets within the utility facilities, 44% of respondents indicated that no such analyses have been performed at their utility. One third of respondents (32%) stated that this type of analysis has been performed for some areas of the plant, and one fifth (20%) said that a formalized Criticality Analysis has been performed for most areas of most facilities.

An effective reliability program includes activities such as regular reviews and adjustments to the maintenance program and standard operating procedures. Nearly half (47%) of applicable respondents indicated that some asset failures and performance shortfalls are reviewed to identify cause and to then develop appropriate failure management policies, and 29% indicated that asset failures and performance shortfalls are reviewed for critical equipment.

Reviews of asset maintenance programs appear to be a common task among respondents, as over half (52%) indicated that some of these programs are reviewed on a regular basis. About 22% indicated that these programs are reviewed on a regular basis for critical equipment, and 12% said that all of these programs are reviewed on a regular basis. When maintenance programs are developed, over half (52%) of applicable respondents stated that some of the process steps used to develop the maintenance programs are formally documented. Nearly one quarter (24%) said that none of these steps are documented.

Earlier, it was noted that reliability is among the top three most serious asset maintenance issues facing utility management. Given the severity of this issue it is no surprise that there is an emergent awareness among senior management with respect to their commitment to reliability initiatives. More specifically, 39% of respondents feel they have strong senior management commitment towards reliability improvement initiatives, and 29% feel management is aware of these initiatives. When asked whether or not their utility has any formalized strategic programs in place to improve reliability of the utility's physical assets, 30% of respondents indicated that their utility is starting a formal program to improve asset reliability, and one quarter (25%) said that there are initial discussions and plans to commence such a program.

Approximately three quarters (74%) of respondents said that their utilities do not have a reliability group in place at their utility. One fifth (21%) of respondents said that an engineer on a part-time basis fulfills the reliability group function.

Utilities generally appear to be making an attempt at keeping up-to-date with new initiatives and industry standards. More specifically, 37% of respondents indicated that their utility takes part in some benchmarking and networking activities and 27% said that their utility belongs to organizations.

### Reliability-Centered Maintenance (RCM)

Of the 177 valid utility responses, 42 went through the section on RCM. [Participation in this section was limited to utilities "where RCM has been used in a significant manner"].

***Key Points on RCM From Utility Survey***

The majority of water utilities use OEM recommendations to formulate maintenance strategy for critical and non-critical assets. RCM is not yet widely used by the water utilities industry (only 10% of the 177 respondents responded to the majority of the RCM questions.)

83% of 18 utilities recognized a positive payback on the RCM effort, with five utilities stating the payback was within 3 months to two years.

The major benefits from the RCM program were stated to be:

- Knowledge Retention
- Asset and System Knowledge
- Improved maintenance work completion
- Reduced maintenance cost
- Operations and Maintenance Co-operation

***Other Salient Data Points on RCM in Utility Survey***

Respondents from those utilities that use RCM were asked to indicate the extent to which RCM has benefited the utility. Knowledge retention for the utility was the most commonly mentioned benefit (60%). Knowledge of assets / systems was another commonly mentioned benefit (58%) along with improved completion rate on maintenance tasks, and reduced maintenance cost (54%). Regarding reliability, 16 % saw significant improvement and 32% saw improvement in some cases.

Benefits were surveyed for thirteen variables. The sample of respondents on benefits ranged from 23 to 25. Across the thirteen variables surveyed, the majority of respondents saw benefit raging from marginal to significant, with approximately one-third seeing no benefit.

In terms of their importance to the utility, respondents ranked reliability as the most important benefit, followed by cost reduction and then knowledge retention.

Derivation and calculation of required frequencies for failure finding tasks for hidden failures is performed in some RCM methodologies (32%). Fully 40% weren't sure whether or not the RCM Methodology used includes this type of calculation.

Respondents were also asked to state their level of awareness with SAE Standard JA1011, which defines standards for acceptable RCM processes. Among applicable respondents, only 12% are familiar with it, and 14% have heard of it, but are not familiar with the details.

**Potential Benefits From Use of RCM Based on Survey Results**

***Duty-Standby Strategies***

Utilities capacity was less than demand for 97% of the utilities surveyed (i.e. the utilities had some excess capacity). The RCM process uses duty-standby strategies to extend equipment life and improve asset reliability. The excess capacity of utilities in normal load conditions, can be leveraged to provide maintenance cost savings and improve equipment reliability by using optimized duty-standby strategies.

*Reduction of Emergency/Breakdown Work*

The failure of assets that affect plant operations that must be rectified immediately (emergency work) exceeds 10% of all work at the majority of utilities. The industry benchmark for emergency work is <5%. The use of RCM can significantly improve equipment availability and assist in improving plant performance by the use of standby-duty operation, improving the preventive maintenance of asset components by the use of condition monitoring techniques and the elimination of counter-productive, intrusive tasks and identifying appropriate maintenance tasks/functional checks for hidden failures.

*OEM-Recommended Maintenance Programs Implies Opportunity for Optimization*

The preventive maintenance programs for approximately half of the utilities surveyed are based on OEM recommendations. The use of OEM-based maintenance programs does not account for the operating context of the equipment in use. The operating context of each plant's equipment (assets) is reviewed during the RCM analysis process and considers the actual operating conditions such as flow, pressure, temperature and process fluid used by each component and its effect on the optimal maintenance required by that asset in those conditions. For example, a sump pump that operates infrequently and is backed up by standby unit may be run to failure, while a critical raw waste water pump that runs 24/7 may require extensive condition monitoring to optimize its availability and avoid consequences of unexpected failure. Plainly the OEM recommendations cannot fit every operating context – even if they have been compiled based on Reliability best practice and RCM concepts (which is frequently not the case). One size does not fit all. Thus this finding that over 50% of maintenance programs are based on OEM recommendations implies that there is much scope for optimization of maintenance programs using RCM.

*Lean Optimized PM Program Leading to High Completion Rates*

The survey results indicate that the preventive maintenance tasks at water utilities are scheduled approximately half the time (48%) and 59% of these tasks are completed. The correct scheduling and completion of preventive maintenance tasks is important to ensure equipment reliability and availability. The establishment and completion of "PMs" is one of the basic building blocks of an asset management program. The industry benchmark for completion of PM tasks is 90%. RCM can improve the preventive maintenance program performance by eliminating low-value PMs, correcting the frequencies on PMs which are set or based on incorrect premise and ensuring that only required PMs based on sound concepts remain. Implementing RCM based on the recommendations laid out in this survey also esnrues that maintenance and operations personnel associated with the asset take ownership of the PM program. Thus RCM programs lead to high completion rates by leaning the PM program to an optimized set of effective tasks while simultaneously building ownership of the program with the people that do the PM tasks and operate the assets.

## Other Asset Management Issues

### *Asset Tagging*

The first step in any asset management program is to identify all assets to be managed/maintained at the utility. Asset tagging using a unique identifier is required to do this to link information to and track activity against that asset. Two thirds of respondents stated that some critical assets were tagged and that much work needed to be done to complete tagging.

### *CMMS System*

Implementation of a computerized maintenance management system (CMMS) is often the next step in the development of an asset management program. CMMS are often not used for small utilities due to the lesser number of assets and size of maintenance staff. The majority of utilities(54%) use a CMMS. Those using CMMS include > 70% of assets in the CMMS database. The average use of the CMMS was five years.

### *Asset Criticality Rating*

The criticality of assets was not established for approximately half of all utilities (44%). The establishment of asset criticality is an important step in work prioritization, budgeting and implementation of an asset management program.

# CHAPTER 1
# INTRODUCTION

## INTRODUCTION

An online survey was administered between April 2004, and January 2005, to water utilities. The survey focused on a maintenance program development process used on physical assets. This process is formally known as Reliability-Centered Maintenance (RCM), and is defined as *a process for deriving the optimal maintenance program for physical assets according to their current user requirements*. This definition was provided to respondents via the Information on the RCM Research Project webpage, which preceded an online survey. A total of 177 respondents completed a valid response to the survey and the results are presented in this report.

The AWWA Research Foundation is interested in specifically understanding how Reliability-Centered Maintenance (RCM) is used in the water industry (namely by water utilities) and the benefits achieved from the application of this process. With increasing regulatory and environmental constraints, tighter budgets, and aging equipment, RCM has emerged as a crucial process in the water industry. Originally developed for the air transport industry, RCM significantly improves safety and reliability, and reduces maintenance costs through the effective maintenance management of physical assets. Those in the water industry are exploiting the benefits realized in the air transport industry by adhering to the principles and processes of RCM.

### Objectives

The objectives of this research are to understand how utilities are applying the RCM process within the water industry, determine the benefits achieved as a result of the application of the RCM process, and to convey information related to the applicability and value of RCM to other stakeholders in the water industry.

# CHAPTER 2
# METHODS AND MATERIALS

An online survey was developed by New Dimension Solutions and was administered to water utilities across the United States as well as in Canada. A total of 177 surveys were completed between April 2004 and January 2005. A single survey completion was accepted from each utility to eliminate redundant responses.

The maximum margin of error at the 95% confidence level associated with the overall sample size of 177 is ± 7.4%.

Caution should be exercised when interpreting results from Reliability – Centered Maintenance section due to the small sample sizes. Values in this section should be treated as directional in nature.

# CHAPTER 3
# RESULTS AND DISCUSSION

## UTILITY OPERATIONS DATA

### *Utility Size – Maximum Day Load Capacity of Respondents*

Respondents represent utilities with varying capacities. Approximately 23% of respondents state that the maximum day load capacity of the utility that they represent is less than or equal to 10 million gallons per day [MGD]. One-fifth (21%) state that the maximum day load capacity of their utility is between 11 and 25 MGD, 19% between 26 and 50 MGD, 13% between 51 and 100 MGD, and 17% between 101 and 300 MGD. About 8% of respondents indicate that the maximum day load capacity of their utility is greater than 300 MGD. The average maximum day load capacity is 101 MGD (*Source: QC2A*)* (Table B.1).

### *Utility Ownership of Responding Utilities*

With respect to the ownership of the utility they represent, over three quarters (77%) of all respondents say that their utility is city or municipality-owned. Special district was a distant second at 7%, followed by investor-owned (6%), county owned (3%) and state-owned (2%) (*Source: QC7*) (Table B.2).

### *Parties Performing Maintenance of Utility Assets*

Respondents were asked to estimate the proportion of their facilities maintained by various parties. Results show that utility facilities are in large part maintained by the utility's employees. Respondents on average indicate that the utility's own employees maintain 85% of facilities. On average, approximately 9% of facilities are maintained by utilities with the assistance of third party contractors, and 6% of facilities are maintained by an outsourced party. On average, only 1% of facilities are maintained by another utility (*Source: QC9*) (Figure B.1).

### *Capacity Level at Which Facilities Are Typically Operated*

When asked how often facilities operate at or near maximum capacity, 40% of all respondents indicated that their utility occasionally operates at this maximum level. One third (32%) of respondents' utilities operate at or near maximum capacity very occasionally and 3% of utilities frequently experience problems matching demand. One quarter (25%) of respondents say that their utility never operates at or near maximum capacity (*Source: QC11*) (Figure B.2).

---

* These references refer to question numbers in the survey questionnaire, which is included elsewhere in this report.

**Table B.1**
**Maximum day load capacity**

| Maximum day load capacity [MGD] | Total (n=172) |
|---|---|
| ≤ 10 | 23% |
| 11 – 25 | 21% |
| 26 – 50 | 19% |
| 51 – 100 | 13% |
| 101 – 300 | 17% |
| > 300 | 8% |

**Table B.2**
**Utility ownership**

| Utility ownership | Total (n=177) |
|---|---|
| City / Municipality Owned | 77% |
| Special District | 7% |
| Investor-Owned | 6% |
| County | 3% |
| State-Owned | 2% |
| Other | 5% |

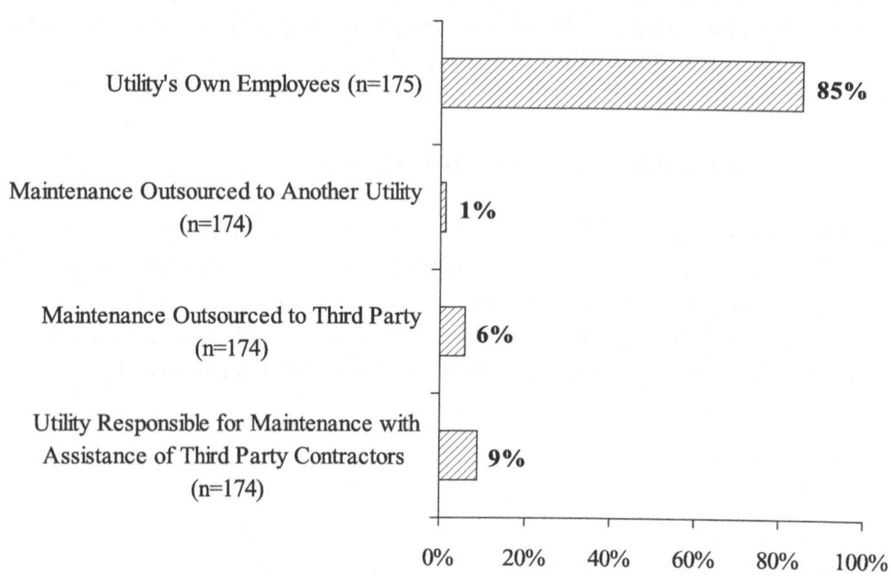


**Figure B.1 Average percentage of facilities maintained by various parties**

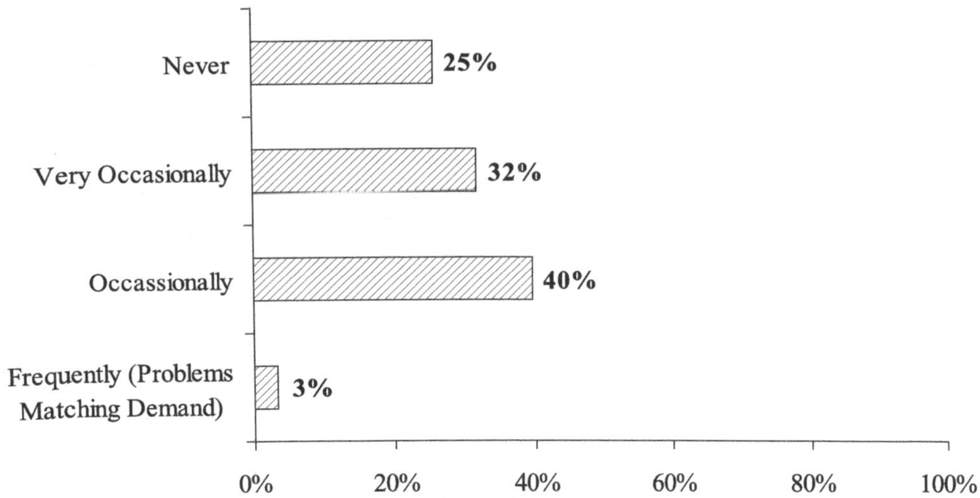


Base: All respondents, n=177

**Figure B.2 Frequency at which facilities operate at or near maximum capacity**

## Current Maintenance / Asset Management Performance and Priority

### *Importance of Asset Management to Responding Utilities*

With respect to the importance of optimal performance and management of the utility's assets to utility management, 57% of respondents say that is becoming an important issue for Utility Management, and 29% indicate that this level of focus is much the same as before. Approximately 13% of respondents say that optimal management and performance of the utility's assets is becoming a very critical issue for Utility Management, and only 1% say that it is decreasing in importance to management (*Source: QD1*) (Figure B.3).

### *Reasons for Any Change in Management Prioritization*

Respondents were asked to provide reasons for any change in management prioritization of physical asset management and performance. Aging facilities and equipment (16%) was the most commonly mentioned reason, followed by better performance and higher efficiency (8%), cost reduction (7%), and budget constraints or restrictions (7%). Other commonly mentioned reasons include concerns about reliability (5%), compliance with regulatory requirements (5%), and higher costs in general (4%) (*Source: QD2A, QD2B*) (Table B.3).

### *Management Satisfaction With Current Utility Physical Asset Management/Maintenance Management*

Management for the most part seems to believe there is room for improvement regarding the current performance of the utilities with respect to physical asset management / maintenance management. More specifically, approximately 54% of respondents feel that the level of management satisfaction with the utility's current performance in terms of physical asset management is

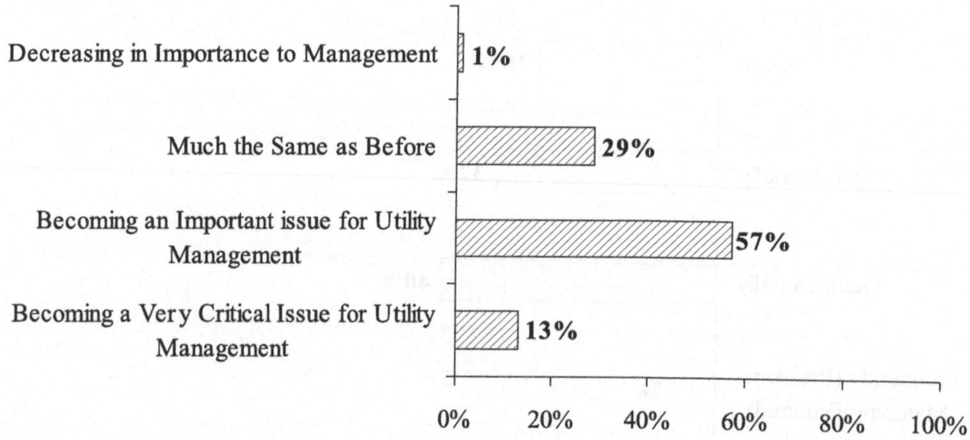


Base: n=161

**Figure B.3 Current importance to utility management of optimal performance and management of the utility's physical assets**

reasonable yet there is still considerable room for improvement. Thirty-five (35%) of respondents say that management is generally satisfied with the utility's current performance in this regard, and 9% say management is not satisfied, as there are major opportunities for improvement. Only 2% feel that management is seriously concerned and that a major improvement must occur (*Source: QD3*) (Figure B.4).

***Most Serious Issues Facing Utilities in Maintaining Physical Assets***

Respondents were asked to specify the most serious issues facing utility management in terms of maintaining its assets. Among the top 3 most serious issues, capital replacement costs (45%) was the most commonly mentioned, followed by maintenance cost (44%), aging facilities (42%), and reliability (40%). Other commonly mentioned issues among the top three included demand catching up to exceed capacity (26%), finding skilled staff (21%), and an aging workforce (16%) (*Source: QD4*) (Table B.4).

***Extent of Utility Adoption of Proactive Maintenance Program***

In terms of how the maintenance department operates, 61% of respondents say that their maintenance department function operates in a steadily improving proactive mode, with some asset systems and facilities still a problem. Approximately 22% of respondents say that their maintenance department function operates with the basics of a proactive program, yet this function is mostly reactive in nature. Eleven percent (11%) of respondents say that their maintenance department function operates in a controlled, proactive environment throughout the utility, and 6% say their function operates in an environment characterized by fire-fighting, with frequent crises and excessive asset downtime (*Source: QD5*) (Figure B.5).

**Table B.3**
**Reasons for any change in management prioritization of physical asset management and performance**

| Top reason | | Included in top two reasons | |
|---|---|---|---|
| Reason | Total (n=123) | Reason | Total * (n=123) |
| Aging Equipment / Aging Facilities / Aging Infrastructure | 16% | Aging Equipment / Aging Facilities / Aging Infrastructure | 21% |
| Better Performance / Higher Efficiency / Optimization (General) | 8% | Better Performance / Higher Efficiency / Optimization (General) | 16% |
| Cost Reduction / Control Costs / Savings | 7% | Cost Reduction / Control Costs / Savings | 10% |
| Budget Constraints / Restrictions / Budgeting (General) | 7% | Compliance w/ Regulatory Requirements / Increased Requirements | 10% |
| Reliability / Improve Reliability / Concerns About Reliability | 5% | Budget Constraints / Restrictions / Budgeting (General) | 9% |
| Compliance w/ Regulatory Requirements / Increased Requirements | 5% | Reliability / Improve Reliability / Concerns About Reliability | 8% |
| Costs (General) / Higher Costs | 4% | Added Demands / Increased Demands / Demands (General) | 7% |
| Added Demands / Increased Demands / Demands (General) | 3% | Costs (General) / Higher Costs | 7% |
| Rates / Control Water Rates / Rate Pressures | 2% | Ensure Assets Function as Intended / Better Asset Control | 4% |
| Extend Life of Facility / Asset | 2% | Rates / Control Water Rates / Rate Pressures | 4% |
| Growing / Expanding System | 2% | Threat of Privatization / Trend Towards Privatization | 3% |
| No Change | 2% | Water Quality / Ensure Water Quality | 3% |
| Change in Management / Restructuring | 2% | Change in Management / Restructuring | 3% |
| Additional Operating Costs / Operating Cost Control | 2% | Growing / Expanding System | 3% |

* Percentages may not sum to 100% due to multiple mention responses and other less commonly mentioned reasons have not been included in this table.

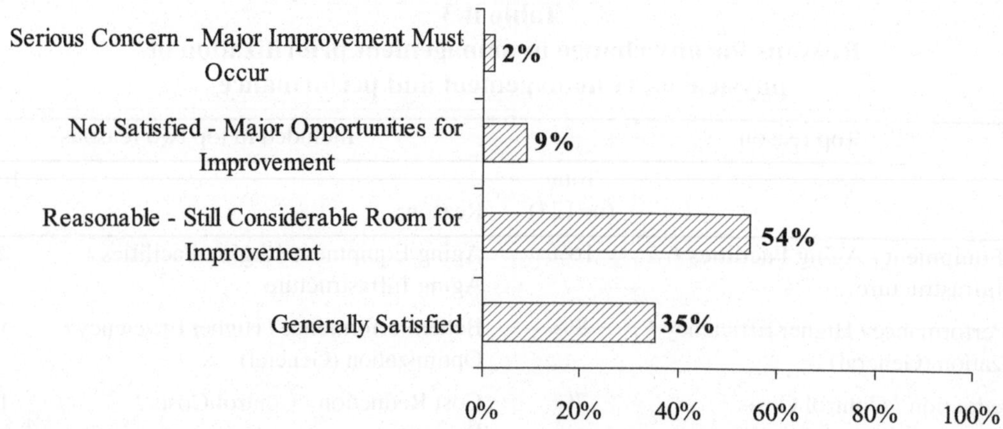


Base: n=161

**Figure B.4 Level of management satisfaction with the current performance of the utilities with respect to physical asset management / maintenance management**

**Table B.4**
**Most serious asset maintenance issues facing utility management**

| Most serious issue | | Included in top three most serious issues | |
|---|---|---|---|
| Asset maintenance issue | Total (n=161) | Asset maintenance issue | Total* (n=161) |
| Reliability | 17% | Capital Replacement Costs | 45% |
| Aging Facilities | 17% | Maintenance Cost | 44% |
| Capital Replacement | 17% | Aging Facilities | 42% |
| Demand Catching Up to Exceed Capacity | 16% | Reliability | 40% |
| Maintenance Cost | 9% | Demand Catching Up to Exceed Capacity | 26% |
| Finding Skilled Staff | 8% | Finding Skilled Staff | 21% |
| Product Quality | 5% | Aging Workforce | 16% |
| Aging Workforce | 3% | Security | 14% |
| Facility in Poor Condition | 2% | Product Quality | 13% |
| Plant Availability | 1% | Distribution Leaks | 9% |
| Environmental Breaches | 1% | Safety Issues | 6% |
| Security | 1% | Facility in Poor Condition | 4% |
| GASB 34 | 1% | Plant Availability | 4% |
| Distribution Leaks | 1% | Environmental Breaches | 4% |
| Other | 2% | GASB 34 | 3% |
| | | Other | 6% |

* Percentages may not sum to 100% due to multiple mention responses.

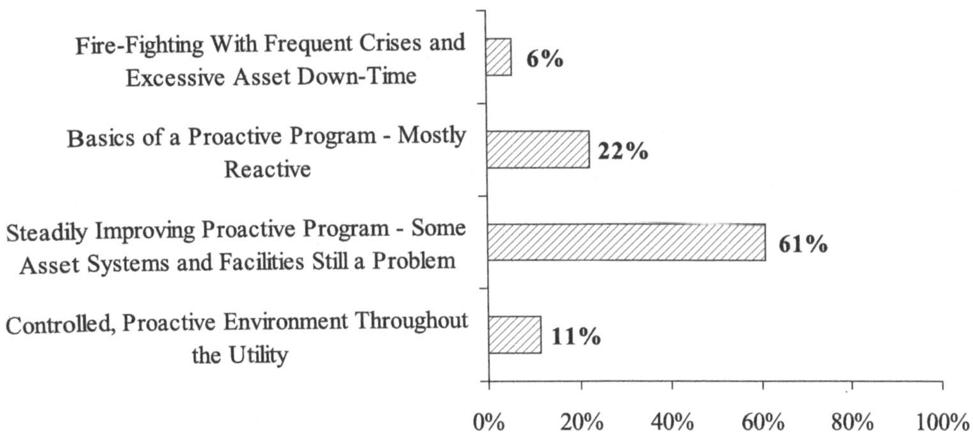


Base: n=161

**Figure B.5 Operation of the maintenance department function**

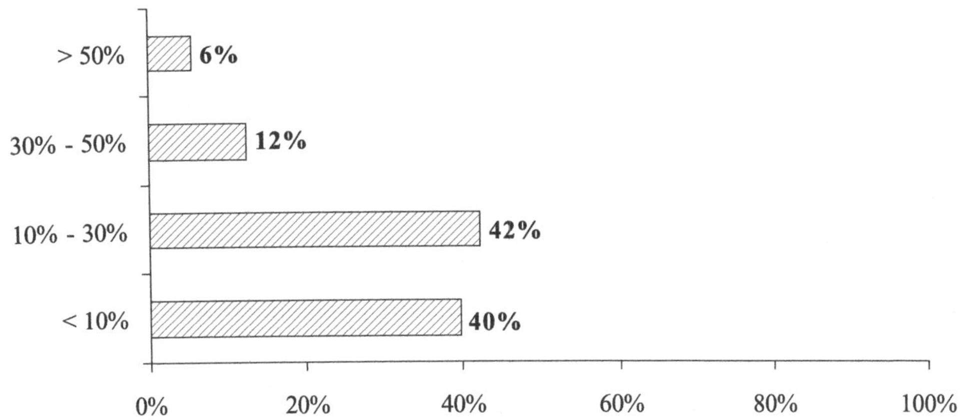


* Base: n=161

**Figure B.6 Percentage of maintenance work in response to plant breakdowns**

***Percentage of Maintenance Work in Response to Asset Breakdown***

In the survey, plant breakdowns are defined as unscheduled failures of an asset that affect plant operations and must be rectified immediately when the asset fails. On average, approximately 20% of maintenance work is in response to plant breakdowns. About 42% of respondents feel that between 10 and 30% of maintenance work is in response to plant breakdowns, 40% say that less than 10% of maintenance work is in response to plant breakdowns, and 12% say that between 30 and 50% of maintenance work is in response to plant breakdowns. Only 6% of respondents feel that over 50% of maintenance work is in response to plant breakdowns. The majority of responses lie at the bottom of the graph, where 82% of respondents indicate that less than 30% of maintenance work is in response to plant breakdowns (Source: *QD6*) (Figure B.6).

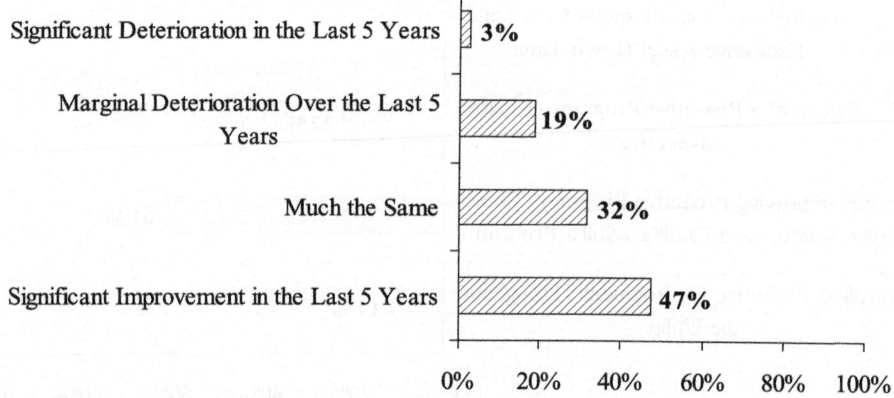


Base: n=161

**Figure B.7 Condition of the utility's facilities / physical assets**

***Trend in Asset Condition at Utilities***

Respondents were asked to specify the condition of the facilities or physical assets of the utilities that they represent. Nearly half of respondents (47%) feel that the condition of the utility's facilities or physical assets has significantly improved over the last five years. One third (32%) of respondents feel that the condition of the utility is much the same, and one fifth (19%) feel that there has been marginal deterioration of these assets over the past five years. Only 3% feel that there has been significant deterioration within the past five years (Source: *QD7*) (Figure B.7).

***Percentage Completion of Fixed Interval (Preventive) Maintenance Tasks***

In terms of fixed-interval maintenance or preventive maintenance tasks, the average percentage of real completion on these tasks currently issued is approximately 59% (Source: *QD8*).

***Credibility and Ownership of Fixed Interval (Preventive) Maintenance Program***

For the fixed-interval / preventive maintenance tasks currently issued, respondents were asked to indicate the level of ownership, credibility and belief that the maintenance work force exhibits. More specifically, half of respondents (51%) feel that the maintenance work force has a reasonable level of acceptance and occasionally provides some feedback on how to improve the fixed-interval tasks currently issued. Approximately 29% say that the maintenance workforce owns the effectiveness of the fixed-interval tasks, and believes in their effectiveness, and nine percent (9%) of respondents say the maintenance work force has some credibility but many of the tasks are worthless or issued at too high a frequency. Six percent (6%) say that there is a very low level of ownership and credibility for the fixed-interval tasks by the maintenance work force, and another 6% indicate that there is no system in place for fixed-interval tasks (Source: *QD9*) (Figure B.8).

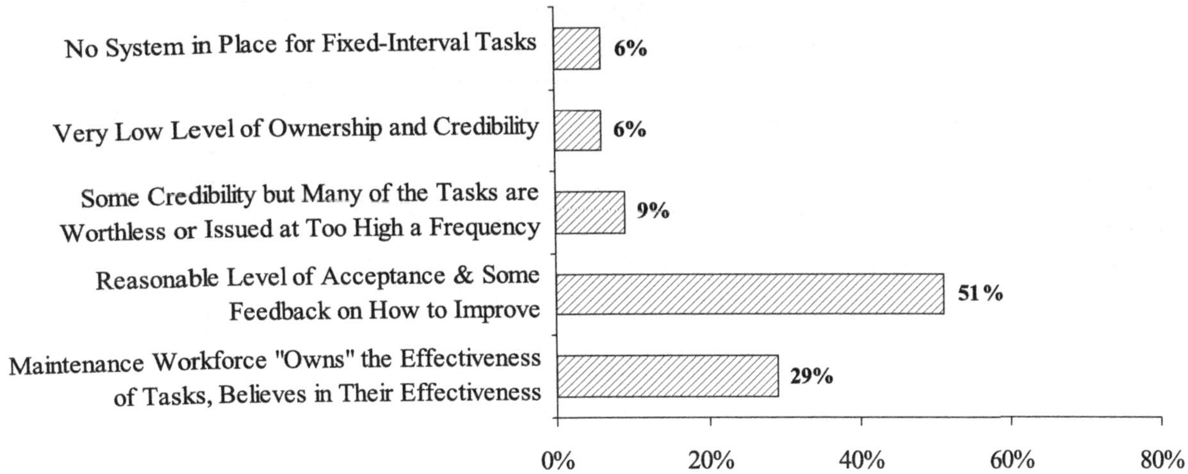


Base: n=161

**Figure B.8 Level of ownership, credibility, and belief that the maintenance work force has in the fixed-interval tasks currently issued**

### *Usage and Effectiveness Ratings of CMMS/EAM Systems and Systems in Use at Respondent Utilities*

Approximately 54% of all respondents indicated that they have a Computerized Maintenance Management System (CMMS) or an Enterprise Asset Management (EAM) system in place. These respondents were asked to indicate the name of the system used, the approximate utility coverage, years the system has been in place, and the effectiveness rating of the system. Among those respondents representing a utility with such a system, 22% say that they use a Maximo system. Approximately 12% use a DataStream MP2 system, and 5% of respondents use an in-house system (Source: *QD10A1*) (Table B.5).

Among the same base of participants, respondents on average indicated that the approximate utility coverage for their system is 71%. The average period of time that the system has been in place is five and a half years, and the average effectiveness rating assigned to this system is 3.6 out of 5, where a 1 represents 'not used', and 5 represents a rating of 'excellent.' This effectiveness rating of 3.6 falls mid-way between 'some benefit' (3) and 'good – can be improved' (4). (*Source: QD10B1, QD10C1, QD10D1*) (Table B.6).

### *Asset Tagging and Registration in EAM/CMMS System*

With respect to the tagging of physical assets in the utility's facilities and registration within the CMMS / EAM system, 35% of respondents say that the tagging and registering of some critical assets has been completed, yet there is still much more work to be done. Thirty-one percent (31%) indicate that the tagging and registering of most assets has been completed, including all critical assets, with some clean-up work to be done. Another 31% say that no assets at all have been tagged or registered, and only 3% of respondents say 100% of the detailed system registration and complete physical tagging has been completed at their facility (*Source: QD11*) (Figure B.9).

**Table B.5**
**Name of CMMS / EAM system**

| Name of CMMS / EAM system | Total (n=96)* |
|---|---|
| Maximo | 22% |
| DataStream MP2 | 12% |
| In-house | 5% |
| Allmax | 4% |
| Hansen / IMS / 7.7 | 4% |
| Access database/program | 3% |
| GBA / GBA MasterSeries | 3% |
| Synergen | 3% |
| DataStream 7i | 3% |

* Percentages do not sum to 100% as other less commonly mentioned systems have not been included in this table.

**Table B.6**
**Average utility coverage, years system has been in place, and effectiveness rating**

| | Average value |
|---|---|
| Average approximate utility coverage (n=94) | 71% |
| Average number of years system has been in place [years] (n=91) | 5.5 |
| Average effectiveness rating [1 – 5]* (n=94) | 3.6 |

* Average effectiveness rating scale: 1 – Not used; 2 – Ineffective; 3 – Some benefit; 4 – Good - can be improved; 5 – Excellent.

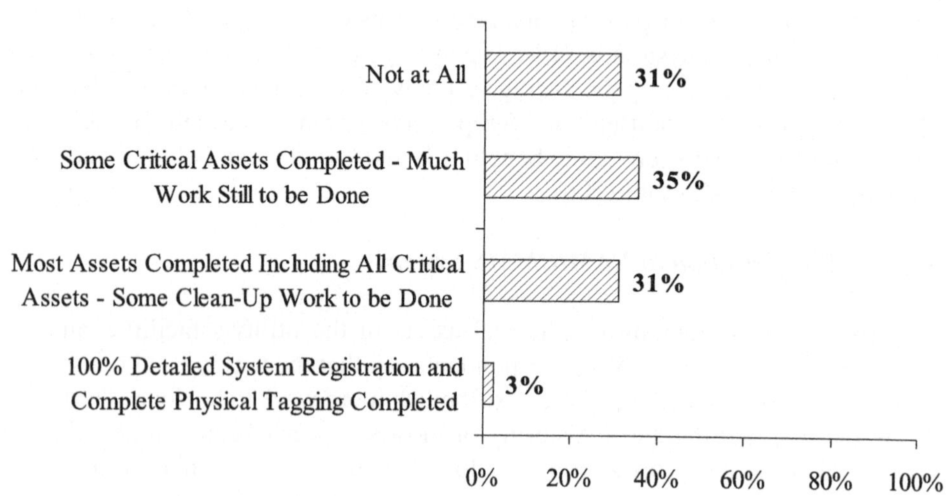


Base: n=161

**Figure B.9 Registration of utility's physical assets in the CMMS / EAM system**

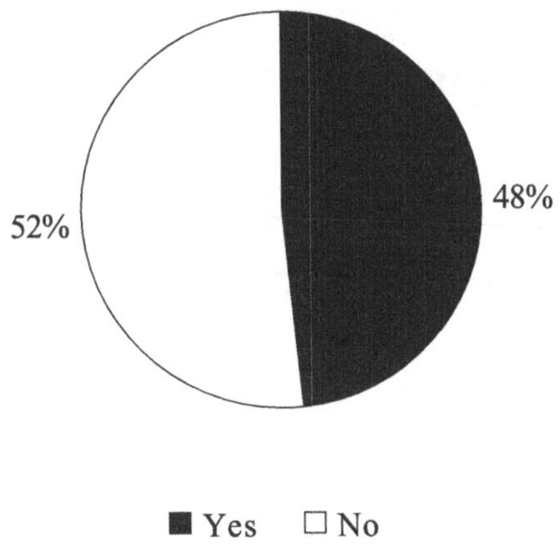


Base: n=161

**Figure B.10 Issuing and scheduling of fixed-interval tasks by the CMMS / EAM system**

**Table B.7**
**Method by which fixed-interval maintenance tasks were developed**

| Development method | Average approximate percentage [%] |
|---|---|
| OEM Recommendations (n=145) | 48 |
| Equipment History (n=144) | 31 |
| Formalized PM Optimization Process (n=140) | 10 |
| RCM (n=132) | 3 |
| Other (n=145) | 8 |

### *Issuing and Scheduling of Fixed-Interval Tasks by CMMS/EAM System*

Respondents were asked whether or not the fixed interval tasks are scheduled and issued by the CMMS / EAM system. Overall, 48% of respondents say that these tasks are scheduled and issued by the CMMS / EAM system (Source: *QD12*) (Figure B.10).

### *Method Used to Develop Fixed-Interval Maintenance Tasks/Program*

To gain a better understanding of the fixed-interval maintenance tasks within their utilities, respondents were asked to indicate the proportion that OEM recommendations, equipment history, formalized PM optimization processes, and RCM contributed to the development of their fixed interval maintenance tasks. Respondents say that fixed-interval maintenance tasks are developed primarily by OEM recommendations (average of 48%), followed by equipment history (average of 31%), formalized PM optimization processes (average of 10%), RCM (average of 3%), and by other means (average of 8%) (*Source: QD13*) (Table B.7).

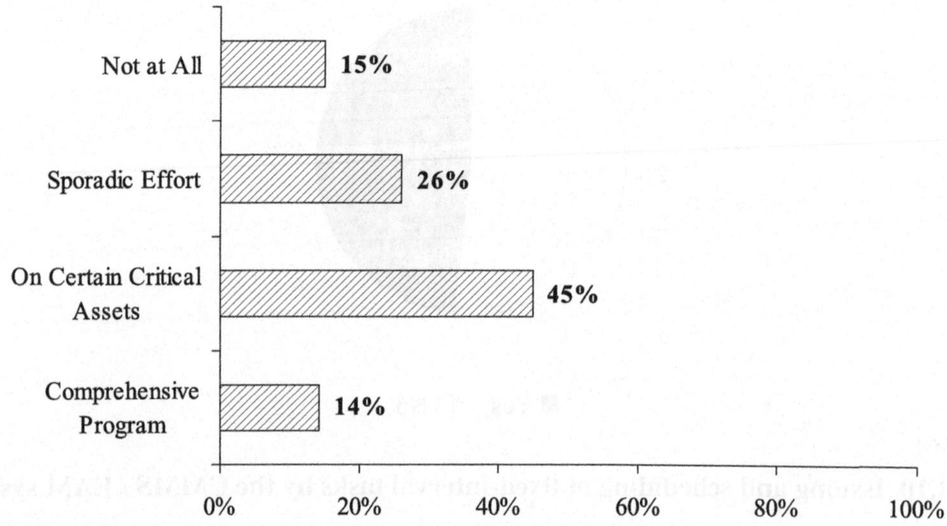


Base: n=147

**Figure B.11 Usage of formal and systematic processes / methods to facilitate development of maintenance programs**

## Maintenance Programs / Failure Management Policies

***Usage of Formal/Systematic Processes for Development of Maintenance Programs***

When asked to indicate whether there are formal and systematic processes / methods in place and utilized at their utility to facilitate development of maintenance programs, approximately 45% of applicable respondents indicated that formal and systematic processes / methods have been utilized at their utility on certain critical assets. One quarter (26%) of respondents say that there is a sporadic effort to implement formal and systematic processes / methods at the utility. Fourteen percent (14%) indicate that there is a comprehensive program in place, and 15% say that there are no formal or systematic processes / methods utilized at the utility to facilitate development of maintenance programs for the utility's assets (*Source: QE1*) (Figure B.11).

***Familiarity With Methods/Techniques for Maintenance Program Formulation***

With respect to the familiarity of the various Maintenance Strategy Formulation techniques, one third (33%) of respondents have a deep working knowledge of OEM, and another 41% have some knowledge of its concepts. Approximately 14% of respondents have a deep working knowledge of PM Optimization, and another 32% have some knowledge of its concepts. Ten percent (10%) of respondents have a deep working knowledge of Reliability-Centered Maintenance, and 26% have some knowledge of its concepts (Source: *QE2*) (Figure B.12).

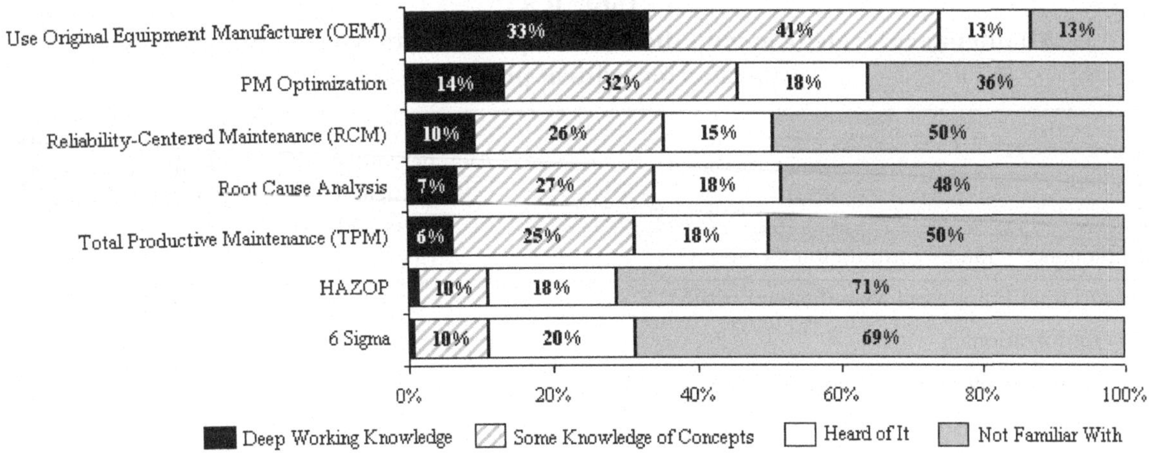


Base: n=147

**Figure B.12 Familiarity with maintenance strategy formulation techniques for developing optimized maintenance programs for physical assets**

### *Maintenance Strategy Formulation Techniques Used for Critical Assets*

Several maintenance strategy formulation techniques are used to develop maintenance programs for *critical assets* within a utility. Among survey participants, the Original Equipment Manufacturer (OEM) program is used on average, on 54% of critical assets at respondents' facilities. PM Optimization is used on 17% of critical assets, and Reliability-Centered Maintenance (RCM) is used on 9% of critical assets. Root Cause Analysis is used on 4% and Total Productive Maintenance is used on 3% of critical assets (*Source: QE3*) (Table B.8).

### *Maintenance Strategy Formulation Techniques Used for Non-Critical Assets*

With respect to maintenance strategy formulation techniques that are used to develop maintenance programs for *non-critical physical assets*, on average, the Original Equipment Manufacturer (OEM) program is used on 56% of these assets at the respondents' facilities. PM Optimization is used on 17% of non-critical assets, and Reliability-Centered Maintenance (RCM) is used on 8% of non-critical assets. Total Productive Maintenance is used on 3% of non-critical assets, and Root Cause Analysis is used on approximately 2% of non-critical assets (*Source: QE4*) (Table B.9).

### *Use of Formalized Maintenance Strategy Formulation During Design-Build Phase of New Assets*

Respondents were also asked if there are any formalized maintenance strategy formulation techniques or reliability optimization methodologies applied in the design-build phase of new assets, or when modifying existing assets. Approximately 44% of respondents indicate that maintenance staff performs some formalized reliability review of the proposed design, and 31% indicate that design engineers and draughtsman try to consider reliability and maintainability during

**Table B.8**
**Maintenance strategy formulation techniques used to develop maintenance programs for critical physical asset**

| Maintenance strategy formulation technique | Avg. percentage of critical assets [%] (n=128) | Pct. of respondents using the various techniques for critical physical assets* (n=177) |
|---|---|---|
| Use Original Equipment Manufacturer (OEM) Program | 54 | 59% |
| PM Optimization | 17 | 32% |
| Reliability-Centered Maintenance (RCM) | 9 | 23% |
| Root Cause Analysis | 4 | 14% |
| Total Productive Maintenance (TPM) | 3 | 10% |
| HAZOP | 1 | 3% |
| 6 Sigma | < 1 | 1% |
| Other | 12 | |
| | Total 100% | |

* A respondent was classified as a user of a specific Maintenance Strategy Formulation Technique if the Percentage of Critical Assets that they entered exceeded 0% for that MSFT.

**Table B.9**
**Maintenance strategy formulation techniques used to develop maintenance programs for non-critical physical assets**

| Maintenance strategy formulation technique | Avg. percentage of non-critical assets [%] (n=129) | Pct. of respondents using the various techniques for non-critical assets* (n=177) |
|---|---|---|
| Use Original Equipment Manufacturer (OEM) Program | 56 | 58% |
| PM Optimization | 17 | 31% |
| Reliability-Centered Maintenance (RCM) | 8 | 18% |
| Total Productive Maintenance (TPM) | 3 | 7% |
| Root Cause Analysis | 2 | 11% |
| HAZOP | < 1 | 2% |
| Other | 12 | |
| | Total 100% | |

* A respondent was classified as a user of a specific Maintenance Strategy Formulation Technique if the Percentage of Non-Critical Assets that they entered exceeded 0% for that MSFT.

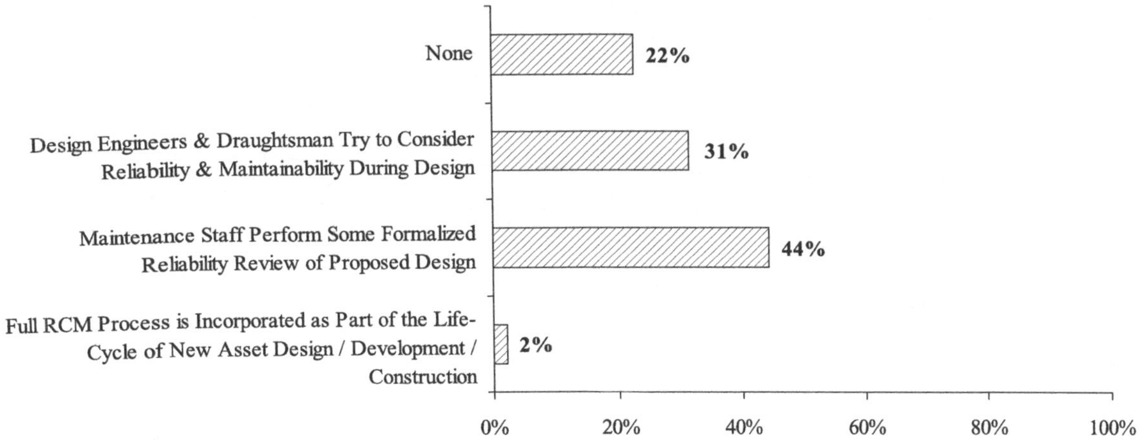


Base: n=147

**Figure B.13 Formalized maintenance strategy formulation techniques or reliability optimization methodologies applied in the design-build phase of new assets**

design. About 22% state that no formalized maintenance strategy formulation techniques or reliability optimization methodologies are applied in the design-build phase or when modifying existing assets. Only 2% of respondents indicate that a full RCM process is incorporated as part of the life-cycle of new asset design / development / construction (*Source: QE5*) (Figure B.13).

### *Formalized Criticality Analysis on Utility Physical Asset Base*

With respect to the determination of a formalized Criticality Analysis* for the technical systems or assets within the utility facilities, 44% of respondents say that no such formalized analysis has been performed for these systems / assets, and one third (32%) say that this type of analysis has been performed for some areas of the plant. One fifth (20%) of respondents indicate that a formalized Criticality Analysis has been performed for most areas of most facilities, and another 4% say that this type of analysis has been determined for all assets at all facilities (*Source: QE6*) (Figure B.14).

### *Review of Asset Failures and Performance Shortfalls to Identify Cause and Adjust/Develop Appropriate Maintenance Program*

An effective reliability program requires ongoing review and adjustment to the maintenance program, / training, / revised standard operating procedures as change occurs or new information comes to light based on real-world performance. When asked whether asset failures and performance shortfalls are reviewed, nearly half (47%) of applicable respondents indicated that some asset failures and performance shortfalls are reviewed to identify cause and to then develop appropriate failure management policies. About 29% indicated that asset failures and performance

* Criticality Analysis is defined as a formal process to set asset priority for maintenance program optimization in terms of failure consequences and likelihood of failure.

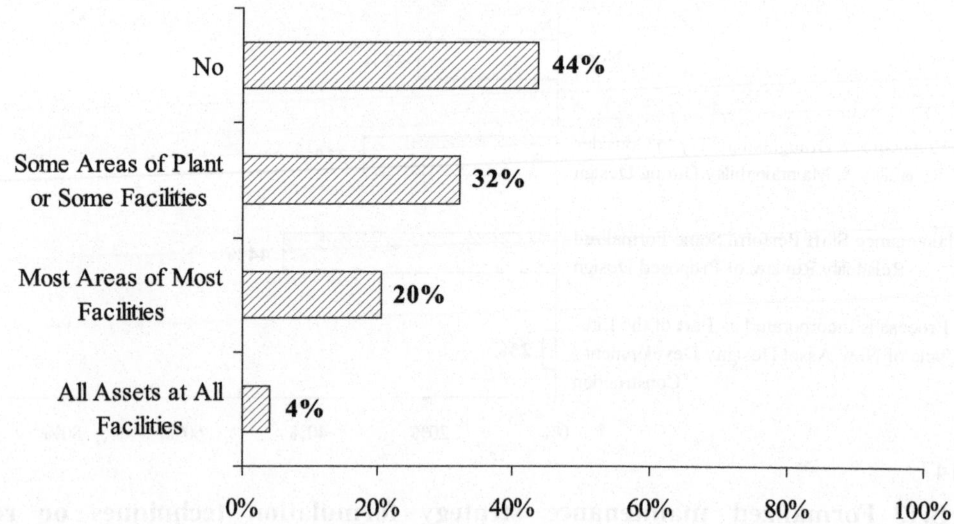


Base: n=147

**Figure B.14 Formalized "criticality analysis" for technical systems / assets in the utility facilities**

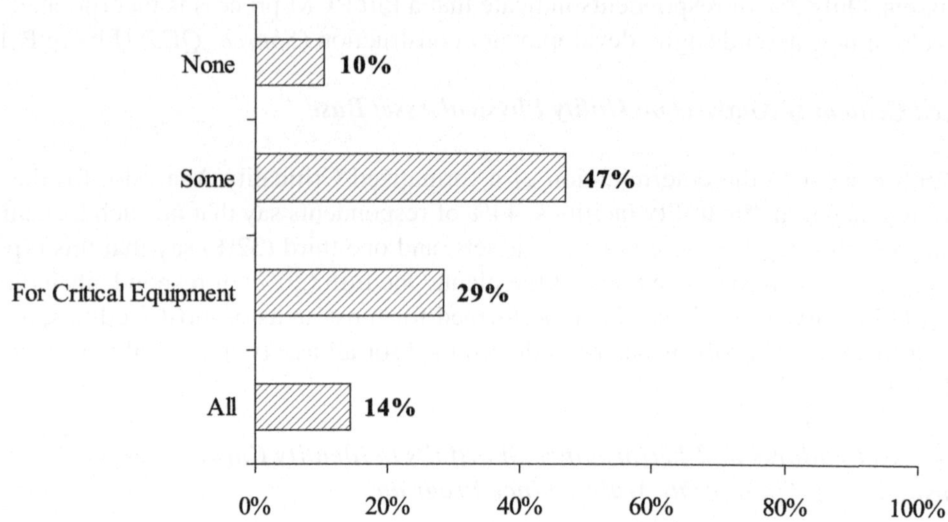


Base: n=147

**Figure B.15 Review of asset failures and performance shortfalls to identify cause and to develop appropriate failure management policies**

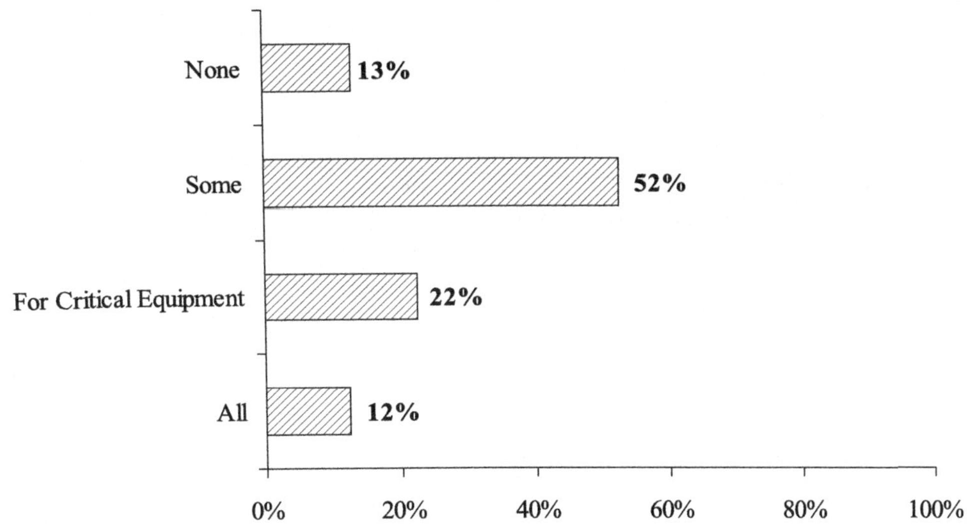


Base: n=147

**Figure B.16 Asset maintenance programs that are reviewed on a regular basis**

shortfalls are reviewed for critical equipment, and 14% said that these reviews are done for all asset failures or shortfalls. One tenth (10%) stated that these types of reviews are not carried out for any of their asset failures or shortfalls (*Source: QE7*) (Figure B.15).

### *Regular Periodic Review of Maintenance Programs*

With respect to the review of asset maintenance programs, over half of respondents (52%) say that some of these programs are reviewed on a regular basis. About 22% indicate that maintenance programs are reviewed on a regular basis for critical equipment, and 12% say that all of these programs are reviewed on a regular basis. Approximately 13% say that no asset maintenance programs are reviewed on a regular basis (*Source: QE8*) (Figure B.16).

### *Responsibilities Assigned to Specific Functions / People in the Utility Organization to Develop Predictive and Preventive Aspects of Asset Maintenance Programs*

Respondents were asked if responsibilities are assigned to specific functions / people in the utility organization, to develop the predictive and preventive aspects of the asset maintenance programs. Approximately 40% of respondents say that responsibilities are assigned to specific functions or people in the utility organization to develop the Predictive and Preventive aspects of asset maintenance programs for some equipment, while 27% say that responsibilities are assigned to develop maintenance programs for all equipment. One fifth of respondents (20%) indicate that these programs are developed for critical equipment, and 14% say that these programs are not developed for any assets (*Source: QE9*) (Figure B.17).

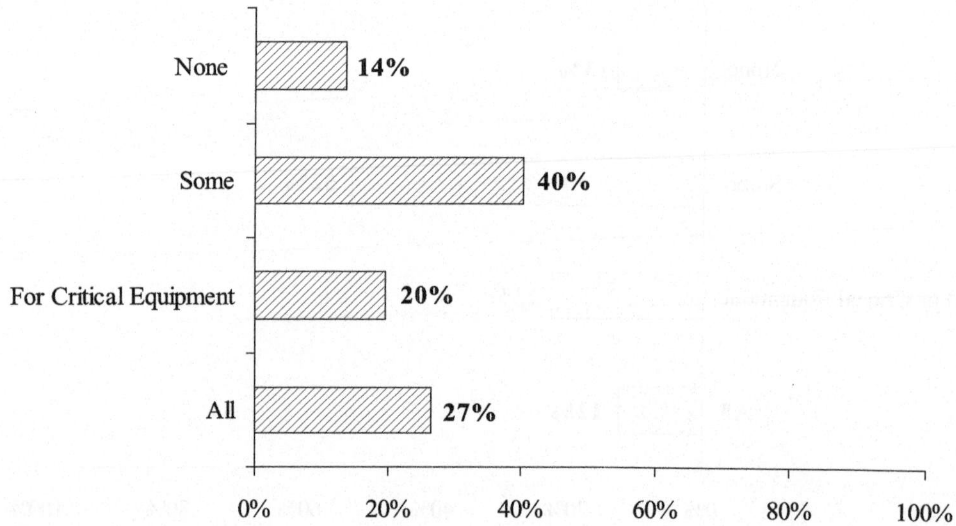


Base: n=147

**Figure B.17 Responsibilities assigned to specific functions / people in the utility organization to develop predictive and preventive aspects of asset maintenance programs**

***Formally Documented Process Steps Used to Develop the Maintenance Programs***

With respect to documenting process steps used to develop maintenance programs, over half (52%) of applicable respondents state that some of the process steps used to develop the maintenance programs are formally documented, whereas about one quarter (24%) say that none of these steps are documented. Fourteen percent (14%) indicate that process steps used to develop maintenance programs are documented only for critical equipment, and 10% say that these steps are documented for all equipment (*Source: QE10*) (Figure B.18).

***Level of Senior Management Commitment Toward Reliability Improvement Initiatives***

When asked to indicate the level of senior management commitment towards reliability improvement initiatives, 39% of respondents said that the utility that they represent has strong senior management commitment toward reliability improvement initiatives, and 29% said that their utility has management awareness of these initiatives. Twenty-seven percent (27%) indicated that there is developing commitment of these improvement initiatives in their utility, and 6% said that there is no senior management commitment towards these initiatives (*Source: QE11*) (Figure B.19).

***Existence of Formalized Strategic Programs to Improve Reliability of the Utility's Physical Assets***

When asked whether or not their utility has any formalized strategic programs in place to improve reliability of the utility's physical assets, 30% of respondents indicated that their utility is starting a formal program to improve asset reliability. About 28% said that there is no formalized strategic program in place to improve reliability, and one quarter (25%) said that there are initial

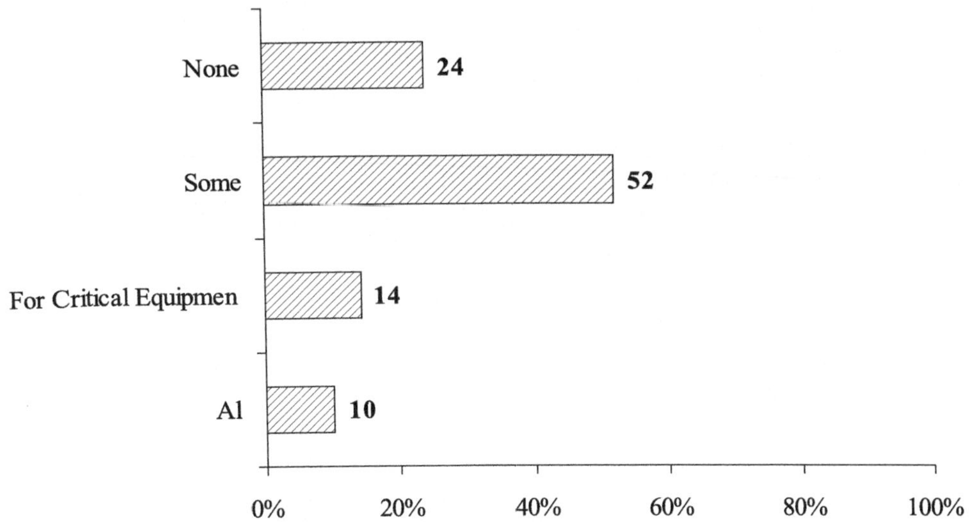


Base: n=147

**Figure B.18 Formally documented process steps used to develop the maintenance programs**

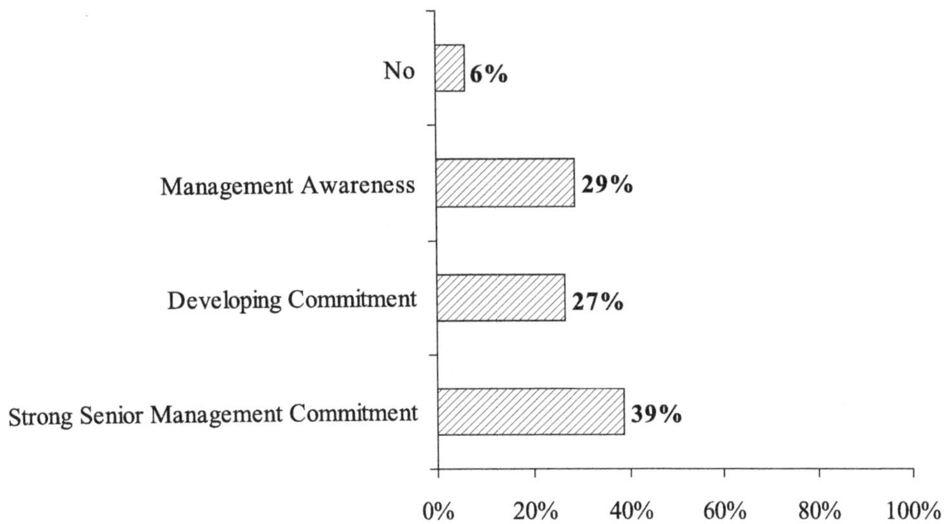


Base: n=147

**Figure B.19 Level of senior management commitment toward reliability improvement initiatives**

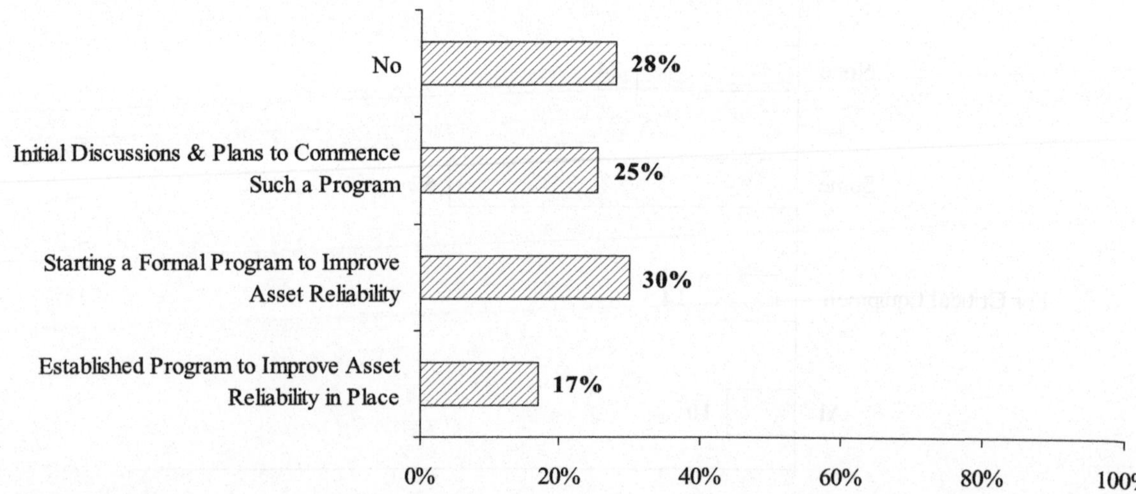


**Figure B.20 Existence of formalized strategic programs to improve reliability of the utility's physical assets**

discussions and plans to commence such a program. Approximately 17% indicated that an established program is in place to improve asset reliability (*Source: QE12*) (Figure B.20).

***Existence of a Reliability Engineering Group or Reliability Group***

Respondents were asked if their utility has a Reliability Engineering Group or Reliability Group in place. Approximately three quarters (74%) of respondents say that their utilities do not have a reliability group in place. One fifth (21%) of respondents say that an engineer on a part-time basis fulfills the reliability group function, and 4% of respondents state that the reliability department fulfills this reliability function. Only 1% of respondents say that a reliability officer fulfills this function within their facility (*Source: QE13*).

Despite the fact that reliability is the single most serious asset maintenance issues facing utility management, there appears to be limited dedicated reliability resources available for Reliability-Centered Maintenance work (Figure B.21).

***Existence of a Formalized Approach to Risk Management***

With respect to a formalized approach to Risk Management, approximately 44% of respondents indicate that their utility has an informal approach in place, and one quarter (24%) say that their utility has a formalized comprehensive Risk Management Policy in place. One fifth of respondents (20%) state there is no such approach in place at their utility, and 12% say that their utility is in the process of formulating a risk management policy (*Source: QE14*) (Figure B.22).

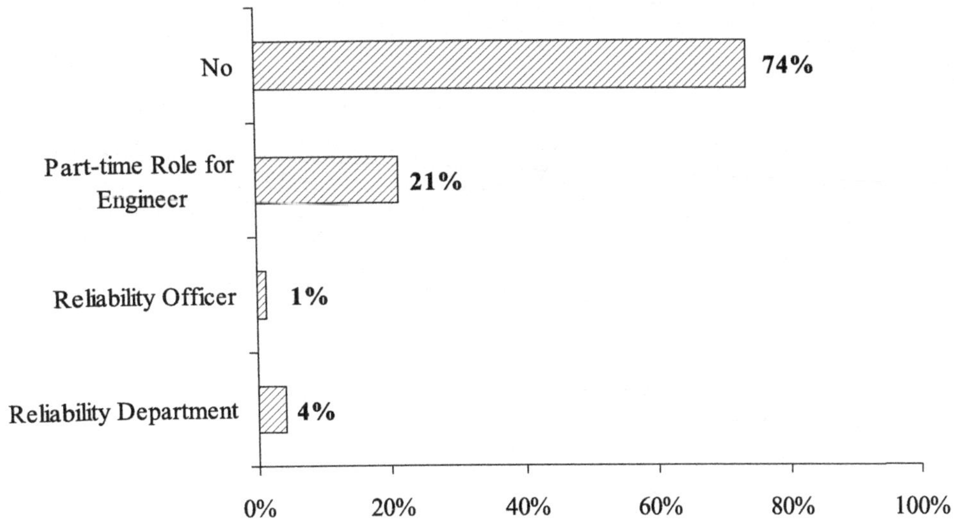


Base: n=147

**Figure B.21 Existence of a reliability engineering group or reliability group**

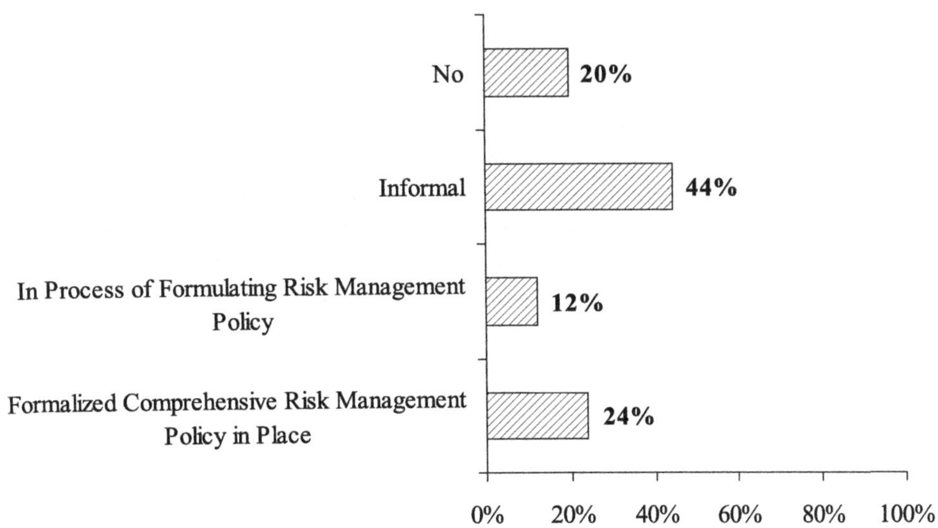


Base: n=147

**Figure B.22 Existence of a formalized approach to risk management**

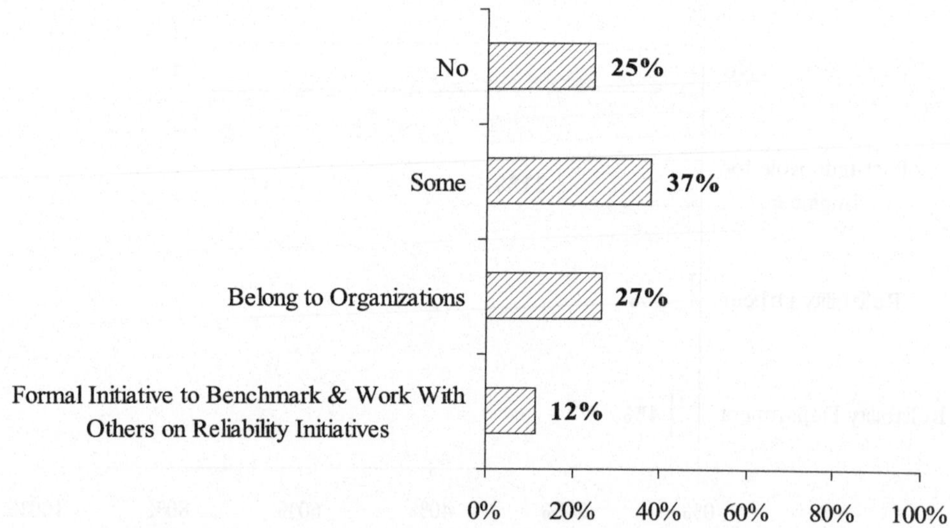


Base: n=147

**Figure B.23 Use of external / internal benchmarking and networking with other utilities to keep up-to-date on new initiatives and industry standards / performance levels / benchmarks**

*Use of External / Internal Benchmarking and Networking With Other Utilities*

To gauge whether or not utilities are attempting to remain up-to-date on new initiatives and industry standards, respondents were asked if their utility takes part in external / internal benchmarking and networking activities with other utilities. Approximately 37% of respondents indicated that their utility takes part in some benchmarking and networking activities to keep up-to-date on new initiatives, and 27% said that their utility belongs to organizations to stay current on new initiatives. One quarter (25%) of respondents indicated that their utility does not take part in these activities, and 12% said there is a formal initiative at their utility to benchmark and work with others on reliability initiatives (*Source: QE15*) (Figure B.23).

*Performance Metrics Defined for the Outputs of Asset Maintenance Programs*

Respondents were also asked if performance metrics are defined for the outputs of asset maintenance programs. These outputs include attributes such as reliability, availability, maintenance costs, work backlog, maintenance cost as % of Replacement Asset Value etc. Forty percent (40%) of respondents say that no performance metrics are defined, whereas one third (33%) say that some of these metrics are defined for the outputs of asset maintenance. About 18% say that some periodic initiatives are defined for the outputs of asset maintenance, and ten percent (10%) indicate that a formal managed program using a CMMS or EAM system is in place (*Source: QE16*) (Figure B.24).

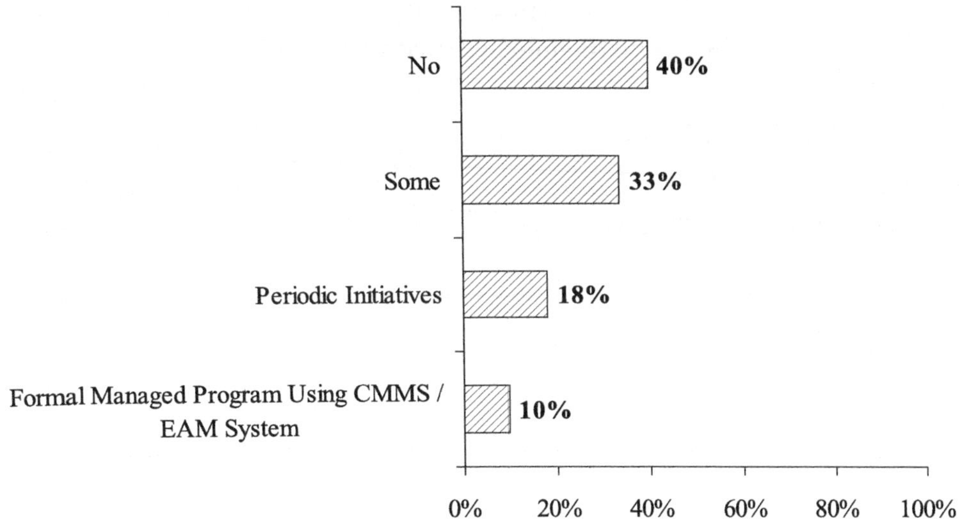


Base: n=147

**Figure B.24 Performance metrics defined for the outputs of asset maintenance programs**

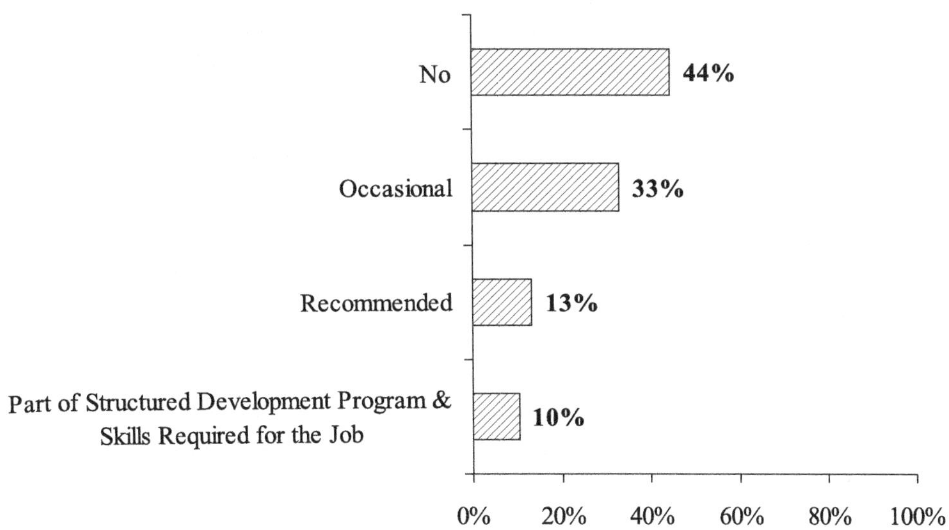


Base: n=147

**Figure B.25 Existence of training plans on any "formalized maintenance strategy formulation techniques" used by the utility for those employees who develop asset maintenance programs**

**Table B.10**
**RCM methodologies used by the utility**

| RCM methodology used | Total (n=39)* |
|---|---|
| No Current Method / None / Not Listed | 59% |
| Aladon RCM (2) | 23% |
| In-house System / We Apply the Concepts | 3% |
| Other | 15% |

* Caution - small sample size

### *Existence of Training Plans on Any "Formalized Maintenance Strategy Formulation Techniques"*

Respondents were asked if there are training plans in place on any formalized maintenance strategy formulation techniques used by the utility for those employees who develop asset maintenance programs. Approximately 44% of respondents state that there are no such training plans in place and one third indicates that there are occasional training plans in place on these formulation techniques. Thirteen percent (13%) state that such training plans are recommended at their utility and 10% say that formalized maintenance strategy formulation techniques are part of a structured development program and are skills required for the job at their utility (*Source: QE17*) (Figure B.25).

## Reliability-Centered Maintenance

[Section only applicable to those Utilities using RCM]

### *RCM Methodology Used by Utility*

Respondents were instructed to complete the RCM section of the survey if RCM was used in a significant manner by their utility. Based on this criterion, 39 out of 177 valid respondents provided information pertaining to the brand of RCM methodology currently or previously used by the utility. Among these respondents, 59% state that there is no specific "brand" or methodology of RCM used, or the "brand" has not been listed. Almost one quarter (23%) of these respondents state that Aladon RCM (2) is used at their utility, 3% use an in-house system, and 15% use other RCM methodologies (*Source: QF1*) (Table B.10).

### *Benefits of Using RCM*

To gain a better understanding of the impact of RCM across utilities, respondents were asked to indicate the extent to which RCM has benefited the utility. See Figure B.26. Among RCM users, 30% indicated that RCM has significantly benefited their facility in terms of knowledge retention for the utility, and 29% indicated that RCM has significantly benefited the utility in terms of: (*Source: QF6*)

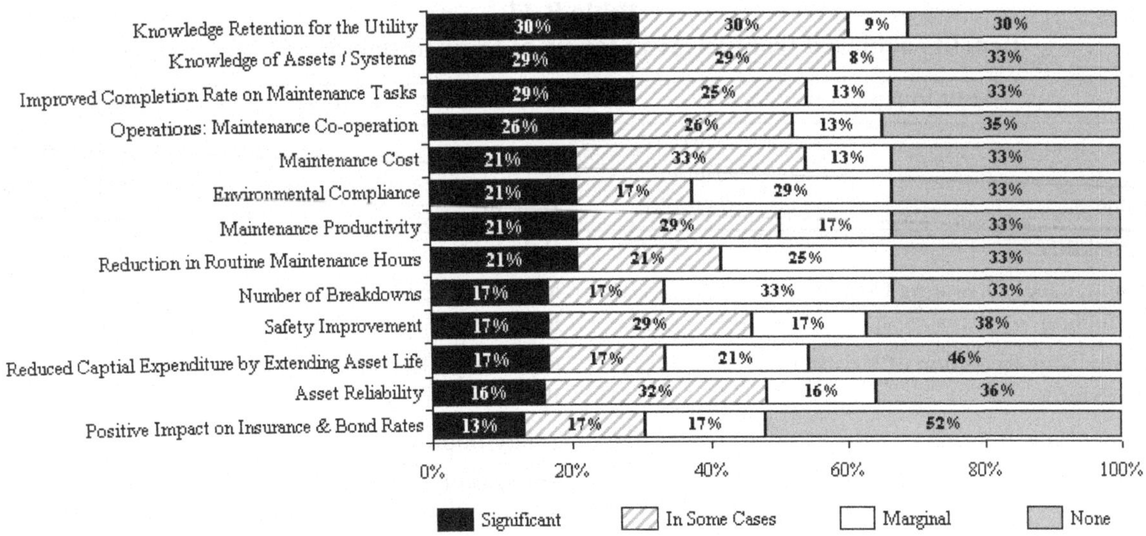


Caution – small sample sizes (n ranging from 23 to 25 for the above 13 attributes)

**Figure B.26 How RCM has benefited the utility**

- Knowledge of assets / systems;
- Improving the completion rate on maintenance tasks.

About one quarter (26%) of these respondents indicated that RCM has significantly benefited their utility in terms of maintenance co-operation with operations, and 21% stated that RCM significantly benefited their utility in terms of:

- Maintenance cost.
- Environmental compliance;
- Maintenance productivity;
- Reduction in routine maintenance hours.

### *Ranking of Benefits Achieved From RCM*

To gauge the importance of RCM, respondents were asked to list the most important benefits achieved from RCM. Among those that use an RCM Methodology, 12 respondents listed reliability as one of the three most important benefits achieved from RCM. Ten (10) respondents mentioned maintenance operating cost reduction, and nine (9) respondents mentioned knowledge retention of existing workforce as one of their three most important benefits that were achieved from RCM (*Source: QF10*) (Table B.11).

**Table B.11**
**Benefits achieved from RCM in terms of their importance to the utility**

| Most important benefit | | Top three most important benefits | |
|---|---|---|---|
| Benefit | No. of respondents (n=20) | Benefit | No. of respondents (n=20) |
| Reliability | 8 | Reliability | 12 |
| Maintenance Operating Cost Reduction | 4 | Maintenance Operating Cost Reduction | 10 |
| Knowledge Retention of Existing Workforce | 3 | Knowledge Retention of Existing Workforce | 9 |
| Safety Improvement | 1 | Safety Improvement | 7 |
| Other | 4 | Improved Environmental Compliance | 3 |
| | | Reduced Capital Expenditure | 3 |
| | | Other | 6 |

Caution – small sample sizes

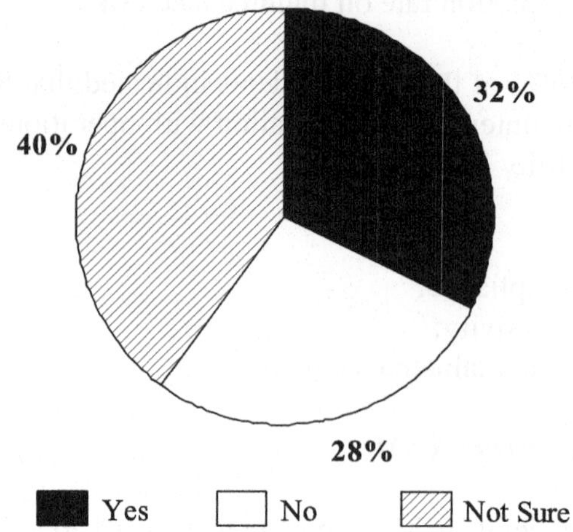


Base: n=25
Caution – small sample size

**Figure B.27 Inclusion of derivation / calculation of required frequencies for failure-finding tasks for hidden failures in the RCM methodology utilized by the utility**

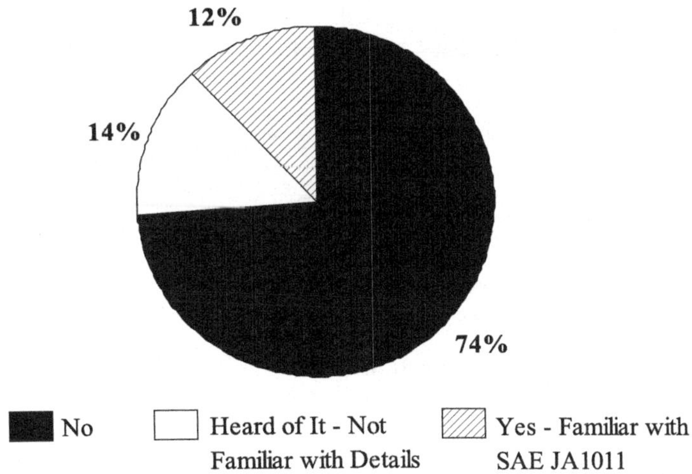


Base: n=42

**Figure B.28 Awareness of the SAE Standard JA1011**

### *Failure Finding Tasks for Hidden Failures*

Derivation and calculation of required frequencies for failure-finding tasks for hidden* failures is performed in some RCM methodologies. Among those that used an RCM Methodology, one third (32%) indicated that the methodology utilized at their utility includes this derivation / calculation of required frequencies, 28% indicated that their methodology does not include the derivation or calculation of these frequencies, and 40% weren't sure whether or not the RCM Methodology used includes this type of calculation (*Source: QF7*) (Figure B.27).

### Familiarity With SAE JA 1011 Standard for Acceptable RCM Processes

Respondents were also asked to state their level of awareness with SAE Standard JA1011, which defines standards for acceptable RCM processes. Among applicable respondents, three quarters (74%) indicated that they are not aware of this standard, 14% have heard of it, but are not familiar with the details, and 12% were familiar with this standard (*Source: QF8*) (Figure B.28).

### *Payback From RCM Projects*

According to the survey, a positive payback is achieved from RCM initiatives if the benefits are greater than the cost of performing RCM analyses. Among applicable respondents, ten (10) indicated that the payback was perceived as positive but was not measured accurately. Three (3) respondents said that there was no payback as the cost of performing the RCM analysis exceeded the benefits. Two respondents said that positive payback was achieved, the first within 6 months to a year, and second was within one to two years (*Source: QF11*) (Figure B.29).

---

* Hidden failures are defined as those failures where the loss of function is not apparent to operators in the normal course of operations. They usually apply to protective functions e.g., alarms, fire extinguishing systems, pressure relief valves, etc.

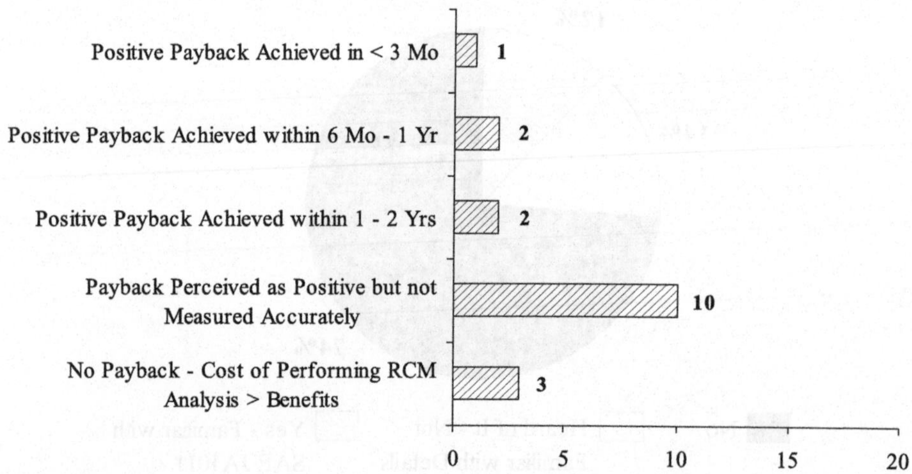


Base: n=18

Caution – small sample size

**Figure B.29 Payback of RCM initiatives pursued by the utility**

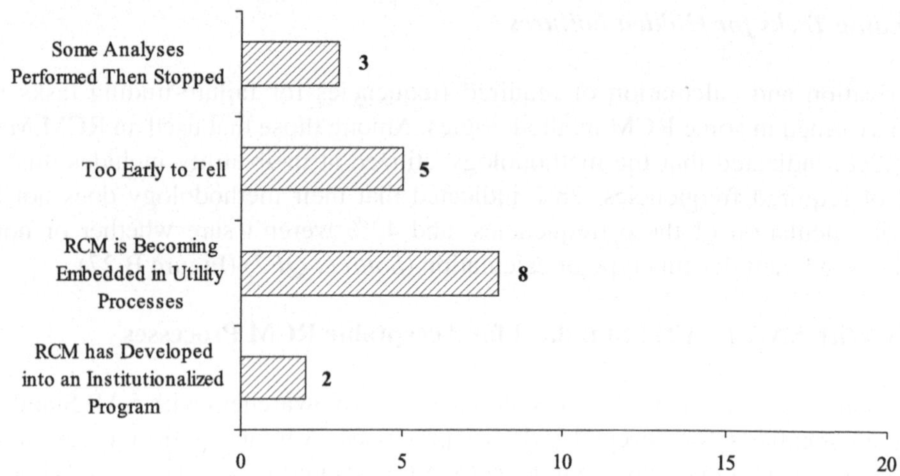


Base: n=18

Caution – small sample size

**Figure B.30 Evolution of RCM within the utility**

***Current Status of RCM Initiatives: Sustainability***

Respondents were asked to specify the state of RCM initiatives at their utility. Approximately eight (8) respondents stated that RCM is becoming embedded in utility processes, and five (5) indicated that it is too early to tell whether or not the RCM initiatives have transitioned from project status into a sustainable program. Three (3) respondents said that some analyses have been performed then stopped, and two (2) have stated that RCM has developed into an institutionalized program (*Source: QF12*) (Figure B.30).

## WATER INDUSTRY QUESTIONNAIRE
## FOR AwwaRF RESEARCH PROJECT:

**Applicability of Reliability-Centered Maintenance in the Water Industry**

**SECTION A: PERSONAL DATA**

QA1. Last Name

QA2. First Name

QA3. E-mail Address

QA4. Telephone number

QA5. Organization (Utility)

QA6. Address 1

QA7. Address 2

QA8. City

QA9. State, Province

QA10. Zip Code [or Postal Code]

QA11. Country

**SECTION B: POSITION DATA**

QB1. Title

QB2. Years with current organization

QB3. Years experience in water industry

QB4. Department

QB5. Management Level

**SECTION C: UTILITY OPERATIONS DATA**

***Drinking Water***

QC1. Number of drinking water treatment facilities (plants)

QC2. Total utility capacity [millions of gallons per day]

QC2A. Max day load capacity [MGD]

QC2B. Average day load capacity [MGD]

QC3. Approx. actual daily average load [MGD]

QC4. Peak demand months of year

QC5. Estimated value of utility physical assets maintained (treatment plants, distribution network fleet, etc.) in million US$--Skip if estimated asset value unknown:

QC6. Does your utility also provide waste-water services? (If yes, an optional section on Waste Water Facilities is included at the end of the survey. Ignore that section if you answer "No.")

❍ Yes
❍ No

QC7. Utility ownership

❍ State-Owned
❍ City or Municipality-Owned
❍ Investor-Owned
❍ Federal-Owned
❍ Other

QC8. Summary of Plant/Facility Information

*Drinking Water Operations*

| | Capacity [MGD] | Approx. age [YRS] | Plant condition 1 – poor 2 – fair 3 – good | % of capacity on normal load | Approx. # of maint. personnel | Approx. # of operations personnel | Level of automation 1 – manual 2 – some 3 – highly automated | SCADA (Y/N) |
|---|---|---|---|---|---|---|---|---|
| Facility 1 | | | | | | | | |
| Facility 2 | | | | | | | | |
| Facility 3 | | | | | | | | |
| Facility 4 | | | | | | | | |
| Facility 5 | | | | | | | | |
| Facility 6 | | | | | | | | |
| Facility 7 | | | | | | | | |
| Facility 8 | | | | | | | | |
| Facility 9 | | | | | | | | |
| Facility 10 | | | | | | | | |

QC9. Utility Maintenance

Insert % of facilities maintained by utility's own employees or other parties as specified. (Entries must total 100%)

| | % of facilities |
|---|---|
| QC9A. Utility's own employees | |
| QC9B. Maintenance outsourced to another utility | |
| QC9C. Maintenance outsourced to third party | |
| QC9D. Utility responsible for maintenance with assistance of third party contractors | |

QC10. Utility Operation
(Entries must total 100%)

| | % of facilities |
|---|---|
| QC10A. Utility's own employees | |
| QC10B. Operations outsourced to another utility | |
| QC10C. Operations completely to third party | |
| QC10D. Operated by utility with assistance of third party contractors | |

QC11. Supply vs. Demand
How often do utility facilities operate at or near maximum capacity?
❍ Never
❍ Very Occasionally
❍ Occasionally
❍ Frequently (problems matching demand)

## SECTION D: CURRENT MAINTENANCE/ASSET MANAGEMENT PERFORMANCE AND PRIORITY

QD1. Is increasing importance being focused on optimal management and performance of the utility's physical assets at this time by utility management?
❍ Decreasing in importance to management
❍ Much the same as before
❍ Becoming an important issue for utility management
❍ Becoming a very critical issue for utility management

QD2. What are reasons for any change in management prioritization of physical asset management and performance?

Reason

QD2A. [ ]

QD2B. [ ]

QD2C. [ ]

QD3. What is the level of Management satisfaction with the utility's current performance in terms of physical asset management/ maintenance management?
❍ Serious concern--major improvement must occur
❍ Not satisfied--major opportunities for improvement
❍ Reasonable--still considerable room for improvement
❍ Generally satisfied

QD4. What are the five most serious issues facing the utility management in terms of maintaining its assets?

QD4A. Most serious

[ ]

QD4B. 2nd most serious

[ ]

QD4C. 3rd most serious

[ ]

QD4D. 4th most serious

[ ]

QD4E. 5th most serious

[ ]

QD4F. Other (description, if applicable)

[ ]

QD5. Does the maintenance department function operate primarily in a reactive mode (reacting to failures when they occur) or in more of a controlled, proactive environment?

❍ Fire-fighting with frequent crises and excessive asset down-time
❍ Basics of a proactive program; mostly reactive
❍ Steadily improving proactive program; some asset systems and facilities still a problem
❍ Controlled, proactive environment throughout the utility

QD6. What approximate percentage of maintenance work (by man-hours) is in response to plant breakdown (unscheduled failure of an asset which affects plant operations and must be rectified immediately when the asset fails)?

❍ >50%
❍ 30-50%
❍ 10-30%
❍ <10%

QD7. Is the condition of the utility's facilities/physical assets generally improving or deteriorating?

❍ Significant deterioration in the last 5 years
❍ Marginal deterioration over the last five years
❍ Much the same
❍ Significant Improvement in the last five years

QD8. What is the average percentage of real completion on fixed-interval maintenance tasks (frequently termed "Preventive Maintenance" or PM) currently issued?

❍ No system exists for fixed-interval (PM) tasks
❍ 20%
❍ 40%
❍ 60%
❍ 80%
❍ >80%

QD9. What level of ownership, credibility and belief does the maintenance work force have for the fixed-interval ("Preventive Maintenance" or PM) tasks currently issued?
❍ No system in place for fixed-interval tasks
❍ Very low level of ownership and credibility
❍ Some credibility but think many of the tasks are worthless or issued at too high a frequency
❍ Reasonable level of acceptance and occasionally some feedback on how to improve the tasks
❍ The Maintenance workforce "owns" the effectiveness of these tasks, believes in their effectiveness, and takes them seriously

QD10. Does the utility have a Computerized Maintenance Management System (CMMS) or Enterprise Asset Management (EAM) System?

| | Name of CMMS/EAM system (e.g. Maximo, Synergen, Indus Passport, Indus MPAC, SAP, Ivara EAM, DataStream, MP2, other) | Approximate utility coverage in % | Years in place | Effectiveness rating 1 – not used 2 – ineffective 3 – some benefit 4 – good – can be improved 5 – excellent |
|---|---|---|---|---|
| System 1 | | | | |
| System 2 | | | | |

QD11. Are the physical assets in the utility's facilities identified with physical tags and name-plates and registered as such in the CMMS/EAM System?
❍ Not at all
❍ Some critical assets completed--much work still to be done
❍ Most assets completed including all critical assets--some clean-up work to be done
❍ 100% detailed system registration and complete physical tagging completed

QD12. Are the fixed-interval tasks scheduled and issued by the CMMS/EAM system?
❍ Yes
❍ No

QD13. How were your fixed-interval maintenance tasks (frequently termed Preventive Maintenance or PM's) developed?
(Entries must total 100%)

| | | Approx % |
|---|---|---|
| QD13A. | OEM Recommendations | |
| QD13B. | Equipment history | |
| QD13C. | Formalized PM Optimization process | |
| QD13D. | RCM | |
| QD13E. | Other | |

## SECTION E: MAINTENANCE PROGRAMS / FAILURE MANAGEMENT POLICIES

QE1. Are there (or have there been) formal and systematic processes/methods in place at the utility to facilitate development of maintenance programs (i.e. to select maintenance tasks) for the utility's assets (equipment)?

❍ Not at all
❍ Sporadic effort
❍ On certain critical assets
❍ Comprehensive program

QE2. Are you familiar with any formalized techniques or methodologies (defined here as "Maintenance Strategy Formulation" techniques) for developing optimized maintenance programs (i.e. performing maintenance task selection) for physical assets?

| | | Not familiar with | Heard of it | Some knowledge of the concepts | Deep working knowledge |
|---|---|---|---|---|---|
| QE2A. | Reliability – Centered Maintenance (RCM) | ❍ | ❍ | ❍ | ❍ |
| QE2B. | Total Productive Maintenance (TPM) | ❍ | ❍ | ❍ | ❍ |
| QE2C. | PM Optimization | ❍ | ❍ | ❍ | ❍ |
| QE2D. | Root Cause Analysis | ❍ | ❍ | ❍ | ❍ |
| QE2E. | 6 Sigma | ❍ | ❍ | ❍ | ❍ |
| QE2F. | HAZOP | ❍ | ❍ | ❍ | ❍ |
| QE2G. | Use Original Equipment Manufacturer (OEM) Program | ❍ | ❍ | ❍ | ❍ |
| QE2H. | Other 1 (description) | | | | |
| QE2I. | Other 2 (description) | | | | |
| QE2J. | Other 3 (description) | | | | |

QE3. What Maintenance Strategy Formulation techniques are used to develop maintenance programs for (i.e. to select maintenance tasks) for critical physical assets at your utility? (Entries must total 100%)

| | | % of critical assets |
|---|---|---|
| QE3A. | Reliability – Centered Maintenance (RCM) | |
| QE3B. | Total Productive Maintenance (TPM) | |
| QE3C. | PM Optimization | |
| QE3D. | Root Cause Analysis | |
| QE3E. | 6 Sigma | |

| | % of critical assets |
|---|---|
| QE3F. HAZOP | |
| QE3G. Use Original Equipment Manufacturer (OEM) program | |
| QE3H. Other 1 (input % and description) | |
| QE3I. Other 2 (input % and description) | |
| QE3J. Other 3 (input % and description) | |

QE4. What Maintenance Strategy Formulation techniques are used to develop maintenance programs (i.e. to select maintenance tasks) for non-critical physical assets? (Make sure total % = 100)

| | % of non-critical assets |
|---|---|
| QE4A. Reliability – Centered Maintenance (RCM) | |
| QE4B. Total Productive Maintenance (TPM) | |
| QE4C. PM Optimization | |
| QE4D. Root Cause Analysis | |
| QE4E. 6 Sigma | |
| QE4F. HAZOP | |
| QE4G. Use Original Equipment Manufacturer (OEM) program | |
| QE4H. Other 1 (input % and description) | |
| QE4I. Other 2 (input % and description) | |
| QE4J. Other 3 (input % and description) | |

QE5. Are any formalized Maintenance Strategy Formulation techniques or reliability optimization methodologies applied in the Design-Build phase of new assets (plants/facilities) or when modifying existing assets?

❍ None
❍ Design engineers and draughtsman try to consider reliability and maintainability during design
❍ Maintenance staff perform some formalized reliability review of proposed design
❍ Full RCM process is incorporated as part of the life-cycle of new asset design/ development /construction

QE6. Has a formalized "Criticality Analysis" (formal process to set asset priority for maintenance program optimization in terms of failure consequences and likelihood of failure) been determined for the technical systems /assets in the utility facilities?

❍ No
❍ Some Areas of Plant or Some Facilities
❍ Most Areas of Most Facilities
❍ All Assets at All Facilities

QE7. Are asset (equipment) failures and performance shortfalls reviewed to identify cause and to then develop appropriate failure management policies (such as adjustments to the maintenance program, training, redesign revised Standard Operating procedures)?
❍ None
❍ Some
❍ For critical equipment
❍ All

QE8. Are asset (equipment) maintenance programs reviewed on a regular basis?
❍ None
❍ Some
❍ For critical equipment
❍ All

QE9. Are responsibilities assigned to specific functions/people in the utility organization to develop the Predictive (PdM) and Preventive (PM) aspects of asset (equipment) maintenance programs?
❍ None
❍ Some
❍ For critical equipment
❍ All

QE10. Are the process steps used to develop the maintenance programs formally documented?
❍ None
❍ Some
❍ For critical equipment
❍ All

QE11. Does the utility have the senior management commitment (CEO, Operations Manager, Director Level) towards reliability improvement initiatives?
❍ No
❍ Management awareness
❍ Developing commitment
❍ Strong senior management commitment

QE12. Does the utility have any formalized strategic program in place to improve reliability of the utility's physical assets?
❍ No
❍ Initial discussions and plans to commence such a program
❍ Starting a formal program to improve asset reliability
❍ Established program to improve asset reliability in place

QE13. Does the utility have a "Reliability Engineering Group" or "Reliability Group" in place?
❍ No
❍ Part-time role for engineer
❍ Reliability Officer
❍ Reliability Department

QE14. Does the utility have a formalized approach to "Risk Management" (i.e. what levels of risk are acceptable in terms of safety, environmental and operational consequences)?
❍ No
❍ Informal
❍ In process of formulating risk management policy
❍ Formalized comprehensive risk management policy in place

QE15. Does the utility use external/ internal benchmarking and networking with other utilities to keep up-to-date on new initiatives and industry standards/performance levels/benchmarks?
❍ No
❍ Some
❍ Belong to Organizations
❍ Formal Initiative to benchmark and work with other interested companies on reliability initiatives

QE16. Are performance metrics defined for the outputs of asset (equipment) maintenance programs? (E.g. Reliability, Availability, Maintenance Cost, Work Backlog, etc.)
❍ No
❍ Some
❍ Periodic initiatives
❍ Formal managed program using CMMS/EAM system

QE17. Are there training plans in place on any "formalized maintenance strategy formulation techniques" used by the utility for those employees who develop asset (equipment) maintenance programs?
❍ No
❍ Occasional
❍ Recommended
❍ Part of structured development program and skills required for the job

**SECTION F: RCM-SPECIFIC QUESTIONS**

[Where RCM has been used in a significant manner by the utility please respond to these questions. If not please leave blank, click Next at the bottom of the page, and continue with the next section.]

QF1. Which "Brand" name(s) of RCM methodology are currently, or have been previously, used by the utility (e.g. Aladon's RCM2; Erin's Streamlined RCM; RCM Turbo etc)? [Please provide information below]

QF2. Name and location of the organization which is marketing/supporting brand/type of RCM methodology used.

QF2A. RCM "BRAND" Name #1

QF2B. Organization Name

[ ]

QF2C. City

[ ]

QF2D. State

[ ]

QF2E. RCM "BRAND" Name #2

[ ]

QF2F. Organization Name

[ ]

QF2G. City

[ ]

QF2H. State

[ ]

QF3. To what extent has RCM been used by the utility?

❍ Several critical systems
❍ A significant number of facility assets/systems
❍ All systems perceived as "critical"
❍ All physical assets

QF4. Utility Rating on RCM Methodology and Training materials used.
Describe each methodology in the blank next to the item number and then select an appropriate Utility Satisfaction Rating on methodology and training materials.

QF4A1. [ ]

QF4A2. **Rating**

❍ Excellent
❍ Good
❍ Average
❍ Poor

QF4B1. [ ]

QF4B2. **Rating**

❍ Excellent
❍ Good
❍ Average
❍ Poor

QF4C1. [ ]

QF4C2. **Rating**

❍ Excellent
❍ Good
❍ Average
❍ Poor

QF4D1. [ ]

QF4D2. **Rating**

❍ Excellent
❍ Good
❍ Average
❍ Poor

QF5. Utility Rating on Organization providing the marketing/support/services for RCM Methodology and Training materials used.

Name each Organization Providing the Marketing/Support/Services for RCM Methodology used by Utility in the blank next to the item number and then select an appropriate Utility Satisfaction Rating.

QF5A1. [ ]

QF5A2. **Rating**

❍ Excellent
❍ Good
❍ Average
❍ Poor

QF5B1. [ ]

QF5B2. **Rating**

❍ Excellent
❍ Good
❍ Average
❍ Poor

QF5C1. [ ]

QF5C2. **Rating**

❍ Excellent
❍ Good
❍ Average
❍ Poor

QF5D1. [ ]

QF5D2. **Rating**

❍ Excellent
❍ Good
❍ Average
❍ Poor

QF6. To what extent has RCM benefited the utility?

| | | None | Marginal | In some cases | Significant |
|---|---|---|---|---|---|
| QF6A. | Asset Reliability | ❍ | ❍ | ❍ | ❍ |
| QF6B. | Number of Breakdowns | ❍ | ❍ | ❍ | ❍ |
| QF6C. | Maintenance Cost | ❍ | ❍ | ❍ | ❍ |
| QF6D. | Safety Improvement | ❍ | ❍ | ❍ | ❍ |
| QF6E. | Environmental Compliance | ❍ | ❍ | ❍ | ❍ |
| QF6F. | Operations: Maintenance Co-operation | ❍ | ❍ | ❍ | ❍ |
| QF6G. | Knowledge of assets / systems: how they work | ❍ | ❍ | ❍ | ❍ |
| QF6H. | Maintenance Productivity | ❍ | ❍ | ❍ | ❍ |
| QF6I. | Knowledge retention for the utility | ❍ | ❍ | ❍ | ❍ |
| QF6J. | Reduction in Routine Maintenance Hours | ❍ | ❍ | ❍ | ❍ |
| QF6K. | Improved completion rate on routine maintenance tasks | ❍ | ❍ | ❍ | ❍ |
| QF6L. | Positive impact on insurance and bond rates | ❍ | ❍ | ❍ | ❍ |
| QF6M. | Reduced capital expenditure by extending asset life | ❍ | ❍ | ❍ | ❍ |

QF7. Did the RCM Methodology utilized by the Utility include Derivation / Calculation of Required Frequencies for Failure-Finding tasks for Hidden Failures (Failures not visible to operators of asset in normal operating context –usually safety and protective functions/devices)?

❍ Yes
❍ No
❍ Not sure

QF8. Are you aware of the SAE Standard JA1011 which defines standards for acceptable RCM Processes?

❍ No
❍ Heard of it - not familiar with details
❍ Yes - familiar with SAE JA1011

QF9. Do you know if the RCM Methodology / processes used by the utility were compliant with this standard?

Input an RCM Process "Brand" Name in each numbered box and then choose the appropriate button.

QF9A1. Brand 1 [ ]

QF9A2. Compliant?

❍ Not known if compliant with SAE JA1011
❍ Not Compliant with SAE JA 1011
❍ Compliant with SAE JA1011

QF9B1. Brand 2 [ ]

QF9B2. Compliant?

❍ Not known if compliant with SAE JA1011

❍ Not Compliant with SAE JA 1011

❍ Compliant with SAE JA1011

QF9C1. Brand 3 [ ]

QF9C2. Compliant?

❍ Not known if compliant with SAE JA1011

❍ Not Compliant with SAE JA 1011

❍ Compliant with SAE JA1011

QF10. Please rank the benefits achieved from RCM in terms of their importance to the utility.

QF10A. Most important [ ]

QF10B. 2nd most important [ ]

QF10C. 3rd most important [ ]

QF10D. 4th most important [ ]

QF10E. Other (description, if applicable) [ ]

QF11. Have the RCM initiatives pursued by the utility achieved a positive payback (benefits greater than cost of performing RCM analyses)?

❍ Positive payback achieved in less than three months
❍ Positive payback achieved within 3 to 6 months
❍ Positive payback achieved within 6 months to 1 year
❍ Positive payback achieved within 1 year to 2 years
❍ Payback perceived as positive but not measured accurately
❍ No payback – cost of performing RCM analyses were greater than benefits

QF12. Have the RCM initiatives transitioned from project status into sustainable program?

❍ Some analyses performed then stopped
❍ Too early to tell
❍ RCM is becoming embedded in utility processes, training programs, etc.
❍ RCM has developed into "institutionalized" program and become part of the way-of-life at the utility

QF13. Please note below any general feedback on Problems/Issues encountered in RCM Initiatives attempted:

[ ]

QF14. Do you perceive that there is a better alternative to RCM in terms of developing optimized maintenance programs for physical assets? If so please comment below:

[ ]

## SECTION G: WASTE WATER TREATMENT

[Applicable only if your utility has Waste Water Facilities. If not, please leave this section blank and click Submit at the bottom of the page.]

QG1. Number of waste water treatment facilities (plants)

QG2. Total utility capacity for waste water treatment [millions of gallons per day]

QG2A. Max day load [MGD]

QG2B. Average day load [MGD]

QG3. Approx. actual daily average load [MGD]

QG4. Peak demand months of year

QG5. Factors influencing waste water loading (e.g. seasonality)

QG6. Approximate value of utility physical assets for waste water (Treatment plants, fleet, etc.) in million US$—skip if not known

QG7. Waste water treatment operations (if applicable)

| | Capacity [MGD] | Approx. age [YRS] | Plant condition 1 – poor 2 – fair 3 – good | % of capacity on normal load | Approx. # of maint. personnel | Approx. # of operations personnel | Level of automation 1 – manual 2 – some 3 – highly automated | SCADA (Y/N) |
|---|---|---|---|---|---|---|---|---|
| Facility 1 | | | | | | | | |
| Facility 2 | | | | | | | | |
| Facility 3 | | | | | | | | |
| Facility 4 | | | | | | | | |
| Facility 5 | | | | | | | | |

## LIST OF PARTICIPATING UTILITIES

We would like to thank the following utilities for their participation in this survey:

Alameda County Flood Control and Water Conservation District Zone 7, Livermore, Calif., Gerald DeWitt
Anchorage Water and Wastewater Utility, Anchorage, Alaska, Kurt Egelhofer
Augusta Utilities Department, Augusta, Ga., Max Hicks
Aurora Utilities, Aurora, Colo., Dana Ehlen
Austin Water Utility, Austin, Texas, Jesus Gomez
Beaches Water Company, St. Leonard, Md., Dennis DiBello
Beatrice Board of Public Works, Beatrice, Neb., Steven Kelley
Bergen County Utilities Authority, Little Ferry, N.J., Eric Andersen
Birmingham Water Works Board, Birmingham, Ala., Billy Shirley
Bloomfield Township, Bloomfield Hills, Mich., Wayne Domine
Board of Water Works, Pueblo, Colo., Terry Book
Brookline Water and Sewer Division, Brookline, Mass., Andrew Pappastergion
Cape May County Municipal Utilities Authority, Cape May Court House, N.J., William W. Cathcart
Carmichael Water District, Carmichael, Calif., Scott Bair
Carson City Development Services, Carson City, Nev., Rit Palmer
Central Arkansas Water, Little Rock, Ark., Stan Benton
City Light and Water, Paragould, Ark., Howerton Lane
City of Akron Water Supply, Kent, Ohio, Jay Strawman
City of Albuquerque Water Utility Department – Water Systems Division, Albuquerque, N.M., Randy Walsh
City of Amarillo, Amarillo, Texas, Dan Coffey
City of Ames Water Plant, Ames, Iowa, Phil Propes
City of Ann Arbor, Ann Arbor, Mich., Sumedh Bahl
City of Arlington, Arlington, Texas, Rick McCleery
City of Ashland Water Treatment Division, Ashland, Ohio, Bruce Wiser
City of Atlanta, Atlanta, Ga., Barry Hamm
City of Billings Public Works, Billings, Mont., Michael Rubich
City of Bloomington, Bloomington, Minn., Terry Neuman
City of Chandler Municipal Utilities, Chandler, Ariz., Wendy Chambers
City of Chelsea Mass. Department of Public Works, Chelsea, Mass., Andrew DeSantis
City of Danville Department of Utilities, Danville, Virginia, Barry Dunkley
City of Davis Public Works, Davis, Calif., Bob Schoech
City of Escondido, Escondido, Calif., Jeff Throop
City of Fairfield - Waterman Treatment Plant, Fairfield, Calif., William Linne
City of Flagstaff Arizona, Flagstaff, Ariz., Randy Pellatz
City of Gillette, Gillette, Wyo., Tara Schultz
City of Gladstone, Gladstone, Ore., Scott Tabor
City of Glendale Arizona Water Utilities, Glendale, Ariz., Rick Scott
City of Golden, Golden, Colo., John McEncroe
City of Grand Junction, Grand Junction, Colo., Rick Brinkman

City of Hagerstown, Hagerstown, Maryland, Chris Bordlemay
City of Independence Water Department, Independence, Mo., Dan Montgomery
City of La Verne, La Verne, Calif., Jerry Mesa
City of Lexington, Lexington, North Carolina, Roger Spach
City of Oceanside, Oceanside, Calif., Mike McGrath
City of Plano - Utility Operations, Plano, Texas, Harry Shearouse
City of Redding Water, Redding, Calif., Michael Conner
City of Richland Water Maintenance Division, Richland, Wash., Scott Siefken
City of Rochester Hills Mich., Rochester Hills, Mich., Roger Rousse
City of Roswell, Roswell, New Mexico, Arthur Torrez
City of Sacramento Utilities Department, Sacramento, Calif., Ronald Adams
City of San Jose, San Jose, Calif., Mary Hoang
City of Santa Ana, Santa Ana, Calif., Thom Coughran
City of Santa Barbara, Santa Barbara, Calif., Jon Peebles
City of Santa Rosa (Utilities Department), Santa Rosa, Calif., Rick Santarini
City of Scottsdale Water / Wastewater, Scottsdale, Ariz., Karl Pape
City of Shreveport - T.L. Amiss Water Treatment Plant, Shreveport, La., Darin Fortenberry
City of St. Marys, St. Marys, Ohio, Dave Sprague
City of Tallahassee Water Utility, Tallahassee, Fla., Jamie Shakar
City of Tempe Water Utilities, Tempe, Ariz., Don Hawkes
City of Thornton, Thornton, Colo., Michael Gay
City of Tulsa, Tulsa, Okla., Clayton Edwards
City of Virginia Beach Dept of Public Utilities, Virginia Beach, Va., Wayne Phelps
City of Waltham, Department of Public Works (was DPW), Water Sewer, Waltham, Mass., Ralph Pecora
City of Westminster, Westminster, Calif., Glen Balogh
Clarksville Gas and Water, Clarksville, Tenn., Kenny Vaughan
Clayton County Water Authority, Morrow, Ga., Herb Etheridge
Columbus Division of Water, Columbus, Ohio, Lynn Kelly
Commissioners of Public Works for the City Of Charleston, Charleston, S. C., Parker Mitchell
Dallas Water Utilities, Dallas, Texas, Charles Stringer
Dare County NC - Water Department, Kill Devil Hills, N. C., Robert Oreskovich
Dearborn Water and Sewerage Division, Dearborn, Mich., James Taylor
Denver Water, Denver, Colo., Robert Lamb
Department of Water Supply - County of Hawaii, Hilo, Hawaii, Daryl Ikeda
Des Moines Water Works, Des Moines, Iowa, Oscarson Douglas
Desert Water Agency, Palm Springs, Calif., Mark Krause
Diablo Water District, Oakley, Calif., Mike Yeraka
East Orange County Water District, Orange, Calif., Jerry Mendzer
EBMUD, Oakland, Calif., Allen Goodson
El Paso Water Utilities, El Paso, Texas, David Brosman
Englewood Utilities, Englewood, Colo., William McCormick
Eugene Water and Electric Board, Springfield, Ore., Doug Wise
Fairfax County Water Authority, Fairfax, Va., Charles M. Murray
Fort Collins Utilities, Fort Collins, Colo., Kevin Gertig
Fort Worth Water Department, Fort Worth, Texas, Robert Arms
Frankfort Electric and Water Plant Board, Frankfort, Ky., Roger Mcdonald

Frankfort Plant Board, Frankfort, Ky., J.H. Sullivan
Golden Heart Utilities, Fairbanks, Alaska, Thomas Rogers
Grand Forks Water, Grand Forks, North Dakota, Hazel Sletten
Greater Cincinnati Water Works, Cincinnati, Ohio, Al Brune
Greater Pine Island Water Association, Bokeelia, Fla., Gary Gissiner
Greenville Utilities Commission, Greenville, North Carolina, Joseph Porter
Gwinnett County Department of Public Utilities, Buford, Ga., Neal Spivey
Highline Water District, Kent, Wash., Thomas Keown
Holland Board of Public Works, Holland, Mich., John Van Uffelen
Honolulu Board of Water Supply, Honolulu, Hawaii, Steve Hayashi
Kitsap PUD, Poulsbo, Wash., Robert Hunter
Lansing Board of Water and Light, Lansing, Mich., Thomas Pietruszka
Lincoln Avenue Water Company, Altadena, Calif., Robert Hayward
Los Angeles Department of Water and Power, Los Angeles, Calif., Stephan Tucker
Manchester Water Works, Manchester, N.H., Russell Johnson
Manitowoc Public Utilities, Manitowoc, Wis., Robert Michaelson
Menasha Electric and Water Utilities, Menasha, Wis., Jerry Sturm
Metropolitan Domestic Water Improvement District, Tucson, Ariz., Steve Shepard
Metropolitan Water District of Salt Lake and Sandy, Sandy, Utah, Steve Stocking
Metropolitan Water District of Southern Calif., La Verne, Calif., Devin Jordan
Miami Dade Water and Sewer, Miami, Fla., Joseph Mazzarese
Milwaukee Water Works, Milwaukee, Wis., Carrie Lewis
Modesto Irrigation District, Modesto, Calif., Patrick Ryan
Mohawk Valley Water Authority, Utica, N.Y., Donald Weimer
Municipal Utility Board / City of Pryor, Pryor, Okla., Gary Pruett
MWRA, Winthrop, Mass., John Colbert
Nevada Irrigation District, Grass Valley, Calif., Robin Lantz
North Andover Water Dept., North Andover, Mass., Dennis Bedrosian
North Clackamas County Water Commission, Ore. City, Ore., Tim Jannsen
Northshore Utility District, Kenmore, Wash., Terry C. Hawley
Oak Creek Water and Sewer Utility, Oak Creek, Wis., Patrick Francis
Ogden City Water Utilities, Ogden, Utah, Danny Stokes
Optechtra, Pretoria, Theuns Koekemoer
Orange County Sanitation District, Fountain Valley, Calif., Simon Watson
Paducah Water, Paducah, Ky., Keith Emmons
Park Water Company, Downey, Calif., Murdoch MacDonald
Pawtucket Water Supply Board, Pawtucket, R.I., Pamela Marchand
Peace River / Manasota Regional Water Supply Authority, Arcadia, Fla., Doug Leath
Plainview Water District, Plainview, N.Y., Paul Granger
Portland Water Bureau, Portland, Ore., Mark Knudson
Poughkeepsies Water Treatment Facility, Poughkeepsie, N.Y., Randy Alstadt
Provo City Water Resources, Provo, Utah, Jan Holley
Public Agency, Fairfield, Calif., Scott Leland
Public Works Commission, Fayetteville, North Carolina, Vernon Madrid
PUD #1 of Clallam County, Port Angeles, Wash., Mike Kitz
Puducah Water Works, Puducah, Ky., Tillman Burnett
Riverhead Water District, Riverhead, N.Y., Mark Conklin

Salt Lake City Department of Public Utilities, Salt Lake City, Utah, Mark Stanley
San Diego County Water Authority, Escondido, Calif., Gary Eaton
San Jose Water Company, San Jose, Calif., Andrew Gere
San Juan Water District, Granite Bay, Calif., Shauna Lorance
Santa Ana River Water Company, Mira Loma, Calif., Arnold Rodriguez
Santa Clara Valley Water District, San Jose, Calif., Alan Zeisbrich
Santa Fe County Utilities Department, Santa Fe, New Mexico, Leonard Quintana
SFPUC Water Supply and Treatment Division, Millbrae, Calif., Paul Bonitz
Sjwd Water District, Lyman, South Carolina, Robert Stewart
Somerville Water and Sewer Enterprise, Somerville, Mass., Barbara Stevens
Sonoma County Water Agency, Santa Rosa, Calif., William Waters, Michael Thompson
Southern Nevada Water System, Boulder City, Nevada, Mark Walters
Tampa Bay Water, Clearwater, Fla., Wayne Smith
The Island Water Association, Sanibel, Fla., John Leiter
The York Water Company, York, Pa., John Strine
Toho Water Authority, Kissimmee, Fla., Brian Wheeler
Town of Estes Park, Estes Park, Colo., Cliff Tedder
Town of Fairview, Fairview, Texas, Duane Anderson
Town of Jupiter Utilities, Jupiter, Fla., Paul Jurczak
Town of West Springfield Department of Public Works (was DPW) Water Division, West Springfield, Mass., Jeffrey Auer
Trenton Water Works, Trenton, N.J., Brandino Cacallori
Truckee Meadows Water Authority, Reno, Nevada, Debra Kaye
United Water, Kirby Pack
United Water, Pine Bluff, Fla., Paul Carlson
United Water Conn. Inc., New Milford, Conn., Kevin Moran
United Water N.Y., West Nyack, N.Y., Richard Ofeldt
Utilities Board City of Sylacauga, Sylacauga, Ala., Darol Russell
Utility Broward County Water and Wastewater Operations Division, Pompano Beach, Fla., Eugene Nicholas
Valencia Heights Water Company, West Covina, Calif., Dave Michalko
Valley of the Moon Water District, El Verano, Calif., Paul Gradolph
Valparaiso Department of Water Works, Valparaiso, Ind., George Brown
Veolia Water Indianapolis, Indianapolis, Ind., Dan Reidy
Village of Downers Grove, Downers Grove, Ill., David Bird
Washington County Department of Water Quality, Williamsport, Md., Kim Bowers
Water Purification Plant, Sioux Falls, S. D., Randy Janssen
Water Replenishment District of Southern Calif., Cerritos, Calif., Paul Fu
Water Utility, Concord, Calif., Doug Young
Wauwatosa Water Utility, Wauwatosa, Wis., Jim Wojcehowicz
West Des Moines Water Works, West Des Moines, Iowa, John Luther
Westchester Joint Water Works, Mamaroneck, N.Y., Richard Ruge
Western Berks Water Authority, Sinking Spring, Pa., Gary Rhoads
Wichita Water Department, Wichita, Kan., Russell Rose
Wisconsin Rapids Water Works and Lighting Comm., Wis. Rapids, Wis., James Reinolt
Ycua, Ypsilanti, Mich., Jeff Castro

# APPENDIX C
# ASSET CRITICALITY ANALYSIS EXAMPLE PRINT-OUTS

| MWRA/WCC<br><br>Criticality Analysis | MASSACHUSETTS WATER RESOURCES AUTHORITY<br><br>PROCEDURE FOR<br>CRITICALITY ANALYSIS | DOCUMENT #:<br>REVISION: 2<br>DATE: 1/08/03<br>REPLACES:<br>PAGE: 1 OF 9<br>TYPE: Administrative |
|---|---|---|

**1.0 <u>POLICY</u>**

1.1. The Criticality Analysis procedure provides MWRA with a documented process for personnel to complete a criticality analysis of a facility, system, or component (FSC).

**2.0 <u>PURPOSE</u>**

2.1. These procedure details the process used for the MWRA Criticality Teams to establish a Total Criticality Number for an FSC.

2.2. The Total Criticality Number is the product of the <u>Total Consequences Number</u> x <u>Probability of FSC failure</u>.

2.3. The Total Criticality Number shall be used to compare the criticality of FSCs agency wide.

2.4. The Total Criticality Number shall be used to prioritize FSCs for inclusion into the RCM analysis process.

2.5. The Total Consequences Number may be used to prioritize corrective maintenance work orders.

**3.0 <u>RESPONSIBILITIES</u>**

3.1. The Deputy Director, Operations and/or Maintenance has ownership of and overall authority over this procedure.

3.2. The Criticality Team Facilitators from Deer Island Treatment Plant, Metro West, Metro East Water, and Metro East Wastewater shall:

1. Select personnel to participate as team members in the criticality analyses.
2. Develop the list of FSCs to be analyzed.
3. Facilitate the criticality analysis meetings.
4. Issue a report documenting all criticality scoring.

3.3. The Criticality Team Members shall attend the criticality analysis meetings and provide input related to the criticality analysis process.

## 4.0 IMPLEMENTATION

4.1. The Criticality Team Facilitator shall select the facilities and systems that shall be analyzed.

1. Clear system boundaries should be defined for each system.

4.2. The Criticality Team Facilitator shall select the analysis team members and schedule the analysis meeting.

4.3. The Criticality Team Facilitator shall enter the following information in the MS Access Criticality Analysis database:

1. On the Facilities Form, enter the following information for each facility to be analyzed:

   a. MWRA Group Code

      1. MEW – Metro East Water
      2. MEWW – Metro East Wastewater
      3. MW – Metro West
      4. DITP – Deer Island Treatment Plant

   b. Facility Name

   c. Facility Code

2. On the Systems Form, enter the following information for each system to be analyzed:

   a. Name of each system. Note that the facility code will be entered automatically.

   b. Details, system boundaries, etc. related to the system (must click on icon next to System Name field to open up details form).

4.4. The following process shall be used during the analysis meeting to determine the criticality of facilities and systems:

1. The analysis meeting format shall be as follows:

   a. The Facilitator shall have a laptop computer with projector set up with the MS Access Criticality database program displayed.

   b. The Facilitator shall issue hard copy print outs of the Facilities and Systems to the analysis team members (reports are available in the Criticality database).

   c. As decisions are made in the meeting the Facilitator shall document them in the database.

2. The initial effort of the analysis shall be to determine the criticality of each facility on the list.

   a. The Facility consequence determination shall be based on the facility's primary function.

   b. The Facility criticality shall be calculated using the probability of failure of the facility's primary function.

**NOTE**

The facility criticality scoring is to be used only as a guideline for the systems analysis. Because the probability of system failure may differ from that of the facility, the systems criticality scores may be higher or lower than the facility criticality score.

3. The subsequent system scoring shall use the facility consequence score as a guideline or upper threshold for the system consequence score.

   a. Safety / Public Health is the only consequence category that can have a system consequence score higher than the facility consequence score.

4. Upon completion of the facilities analysis and scoring, each listed system and components within each facility shall be analyzed and scored.

4.5. Complete the following steps to calculate the Total Consequence Number and the Total Criticality Number of either a facility or system (refer to Attachment 1, Equipment Criticality Analysis, Consequence / Probability Matrix):

1. The analysis team shall determine the consequence score for each consequence category on the matrix.

   a. Assume the following when defining the consequences of FSC failure:

      1. The entire facility, system, or component is in a failed state.

      2. Highest flow capacity is assumed at the time of failure.

   b. When selecting a score, always use the number values provided on the matrix in each category. DO NOT enter a score that is in between the number values shown.

   c. For the Safety/Public Health and Environmental /Product Quality categories, refer to pages 2 or 3 of Attachment 1 for scoring guidelines.

d. For the Financial Impact category include the following costs:

1. Damage to equipment during a fire.
2. Flood or wastewater damage to private or Authority property.
3. Costs associated with loss of life.

e. The Facilitator shall enter the consequence score in the Criticality database.

f. The Facilitator may enter details related to the consequence determination in the System or Facility details field (click icon to open up memo field).

g. The total consequence number is automatically calculated as the sum of the four consequence scores and displayed in the database.

2. The analysis team shall estimate the probability of failure for each facility or system.

a. Referring to Attachment 1, Equipment Criticality Analysis, Consequence / Probability Matrix, the team shall select the probability score corresponding to the assumed failure rate of the facility or system.

1. The failure rate shall be based on the knowledge and experience of the staff most familiar with the facility or system.
2. If historical failures have been addressed and are not expected to repeat, then the assumed future failure rate should be used.

b. If a multiple failure has to occur in order to have consequences, the probability shall be calculated as the product of the probabilities of the multiple failures (Total probability = Failure A Probability x Failure B Probability) Consider the following examples:

1. The failure of a protective system only has consequences when what it is protecting against occurs, such as a fire alarm system failure only has consequences during a fire. If the probability of a fire alarm system failure is 1 in 10 years (.10 failures per year) and the probability of a fire is 1 in 20 years (.05 fires per year) the probability of having a fire alarm failure during a fire is calculated as 1/10 (.10) x 1/20 (.05) = 1/200 (.005) (this scenario has the probability of occurring once in 200 years). Therefore, a probability of the next highest probability (1 in 150 years) or 0.1 should be used.
2. The failure of a backup or standby system only has consequences when the primary system fails. Therefore, the probability of failure should be calculated as the probability of backup system failure x probability of primary system failure.

3. Failure of an Air Conditioning system will only matter if cooling is required (warm summer days). Therefore, if it is determined that an AC system only has failure consequences on days that the outside temperature is above 80 degrees Fahrenheit, then the probability shall be calculated as the probability of AC failure times the probability of +80 degree days occurring.

c. The Facilitator shall enter the probability score in the Criticality database.

d. The Facilitator may enter details related to the probability determination in the System or Facility details field (click icon to open up memo field).

e. The Total Criticality Number is automatically calculated as the product of the Total Consequence Number x Probability and displayed in the database.

3. Upon completion of each analysis meeting the Facilitator shall backup the Criticality Database on disk.

4.6. Upon completion of all facility and system analyses, the Facilitator shall forward the final Criticality database to the Deer Island Asset Manager for addition of the records to the overall MWRA Criticality database.

4.7. The DITP Asset Manager shall issue a report that details the criticality ranking of all MWRA facilities and systems to the appropriate Operations and Maintenance managers.

## 5.0 <u>ATTACHMENTS</u>

5.1. Attachment 1, Equipment Criticality Analysis, Consequence / Probability Matrix

# Equipment Criticality Analysis, Consequence/Probability Matrix

| Consequence | | | | Probability |
|---|---|---|---|---|
| **Safety/ Public Health** | **Environmental/ Product Quality** | **Customer Service** | **Financial Impact** | **Failures Occur** |
| 40 = Major public health risk or Multiple fatality | 32 = Major environmental damage or product quality impact | 20 = > 1000 customer | 40 = impact > $ 2.5 Million | 10 = daily to < weekly<br>8 = weekly to < monthly<br>7 = monthly to < yearly<br>4 = yearly to < 5 years<br>3 = 1 in 5 to <10 years<br>2 = 1 in 10 to <20 years<br>1 = 1 in 20 to <40 years<br>0.5 = 1 in 40 to <80 years<br>0.25= 1 in 80 to <150 years<br>0.1= 1 in 150 to <300 years<br>0.05 = 1 in 300 and > years |
| 32 = Public health risk or fatality | 24 = Significant environmental damage or product quality impact | 16 = > 500 and < 1000 customers experiences a loss of service | 32 = impact $ 500,000 to $ 2.5 million | |
| 24 = Low public health risk or major serious injury | 16 = Minor environmental impact or product quality impact | 12 = > 100 and < 500 customers experiences a loss of service | 24 = impact $ 100,000 to $ 500,000 | |
| 16 = Minor public health risk or moderate injury | 8 = Minor or no environmental or product quality impact | 8 = 0 to 100 customers dissatisfied | 16 = impact $ 50,000 to $ 100,000 | |
| 0 = None | 0 = None | 0 = None | 0 = None | |

Attachment 1

## Criticality Analysis
## Detailed SAFETY/PUBLIC Health Rating Scale

40 = MAJOR PUBLIC HEALTH RISK OR MULTIPLE FATALITY

1. Multiple fatality > 5
2. Major contamination of the drinking water system causes illness and death.
3. Long term (> 8 hours) Interruption/cessation of water adequate for fire flow pressure and or to meet sanitation needs.
4. Public exposure to raw wastewater that causes widespread illness and/or death.

32 = PUBLIC HEALTH RISK OR FATALITY

1. < 5 fatalities
2. Boil-water order implemented by DEP.
3. Discharges to the atmosphere, sickness from water, or public exposure to raw wastewater that requires widespread treatment by health care professionals.
4. Interruption (between 4-8 hours) of the transmission/distribution of drinking water to a component of the service area resulting in inadequate pressure for fire flow requirements and water to meet sanitation needs.

24 = LOW PUBLIC HEALTH RISK OR MAJOR SERIOUS INJURY

1. Serious injury
2. Adequate water pressure is not provided (less than 35 psig). Enough pressure is provided for limited fire flow and sanitation requirements but does not meet customer expectations.
3. Discharges to the atmosphere, sickness from water, or public exposure to raw wastewater that requires limited treatment by health care professionals.

16 = MINOR PUBLIC HEALTH RISK OR MODERATE INJURY

1. Moderate injury including contusions, lacerations or chronic illness from working environment.
2. Discharges raw wastewater that requires no treatment by health care professionals.

0 = NONE

Attachment 1

**Criticality Analysis**
**Detailed Environmental/Product Quality Consequences Rating Scale**

32 = POTENTIAL FOR MAJOR ENVIRONMENTAL DAMAGE AND OR PRODUCT QUALTIY IMPACT

1. Wastewater process has an upset that results in severe environmental damage (immediate fish kills, etc.)
2. MWRA does not meet minimum Safe Drinking Water Act/Surface Water Treatment Rule standards or MA Drinking Water Standards for pathogens or activation.
3. Major kill of wildlife—generally fish or birds

24 = POTENTIAL FOR SIGNIFICANT ENVIRONMENTAL OR PRODUCT QUALITY IMPACT

1. Wastewater process failure that results in moderate environmental impact
2. MWRA does not meet internal drinking water treatment process objectives for primary disinfection, pH, alkalinity, fluoridation, or DBPs.
3. MWRA exceeds acceptable level of taste and odor of drinking water

16 = POTENTIAL FOR MINOR ENVIRONMENTAL OR PRODUCT QUALITY IMPACT

1. Wastewater or water process failure that results in minor environmental impact.
2. Water Quality conditions cause white water, specks or color problems that cause damage to customer property.
3. Discharges to the watershed or tributary of a drinking water reservoir, storm sewer, sanitary sewer, or waterway, which exceeds regulations or accepted standards and practices.
4. Reportable release of chemicals/toxins to the ground

8 = MINOR OR NO ENVIRONMENTAL OR PRODUCT QUALITY IMPACT

1. Water Quality is impacted by white water, specks or color problems that causes no damage to customer property
2. Releases of process fluids inside buildings or to containment structures which are not physically connected to the drinking water, treatment or distribution system, and/or do not get to natural environment
3. Operation of process equipment without environmental equipment (odor control, etc.) even if there is no permit violations.

0 = NONE

Attachment 1

## Equipment Criticality Analysis, Consequences Rating Scale

| Effort | Safety | Environmental | Operational | | | | Frequency/ Probability |
|---|---|---|---|---|---|---|---|
| | | | Product Quality | Output | Customer Service | Operating Costs | |
| 5 = >60 meetings | 40 = Multiple fatility | 40 = Permanent environmental damage | 8 = (extreme) cannot rework or be sold as seconds. SCRAP | 8 = (extreme) Inability to make up annual production quotas, lost revenue | 8 = (extreme) loss of customer and potential litigation | 8 = (extreme) major loss of revenue, long term reduced earnings $500 000> | 10 = Failures occur daily. |
| 4 = 40 - 60 meetings | 36 = Fatality | 36 = Major incident with external plant effects and temporary environmental damage | 6 = (major) cannot rework, but can be sold as seconds | 6 = (major) cannot make up lost production within the company, have to buy out | 6 = (major) customer experiences downtime or excessive scrap rates, costs charged back to the company | 6 = (major) substantial loss of revenue $100 000> but <$500 000 | 9 = Failures occur weekly. |
| 3 = 20 to 40 meetings | 32 = Major serious injury (disabling) | 28 = Incidents with internal plant impact only | 4 = (moderate) cannot reapply, have to rework | 4 = (moderate) No excess line capacity, but lost production can be made up within the company | 4 = (moderate) customer rejects product or late delivery | 4 = (moderate) major increase in variable costs $10 000 > but <$100 000 | 8 = Failures occur monthly. |
| 2 = 10 to 20 meetings | 24 = Moderate injury (lost time) | 20 = Minor incidents | 2 = (slight) out of spec, can be reapplied | 2 = (slight) can recover lost production through excess plant capacity | 2 = (slight) customer disatisfied but prepared to use | 2 = (slight) minimal increased variable costs ie. overtime, energy etc. <$10 000 | 6 = Failures occur yearly. |
| 1 = <10 meetings | 20 = Minor injury (contusions and lacerations | 0 = No emmissions | 0 = None | 0 = None | 0 = None | 0 = None | 4 = Failures occur 1 in 5 years. |
| | 0 = No injury | | | | | | 2 = Failures occur 1 in 10 years. |
| | | | | | | | 1 = Failures occur less frequently than 10 years. |

# APPENDIX D
# INTRODUCTION TO RCM
# (COURTESY OF ALADON)

## 1. INTRODUCTION

Reliability-centered Maintenance (RCM) finds its roots in work done by the international commercial aviation industry. Driven by the need to improve reliability while containing the cost of maintenance, this industry developed a comprehensive process for deciding what maintenance work is needed to keep aircraft airborne. This process evolved steadily since its early beginnings in 1960.

In 1978, a report[1] was prepared for the US Department of Defense describing the then current state of the process. The report was written by F Stanley Nowlan and Howard Heap of United Airlines, and it was entitled "Reliability-centered Maintenance," or RCM. It formed the basis of the maintenance strategy formulation process called MSG3.[2] MSG3 was first promulgated in 1980, and in slightly modified form, it is used to this day by the international commercial aviation industry. In the early 1980's, RCM as described by Nowlan and Heap also began to be used in industries other than aviation.

It soon became apparent that no other comparable technique exists for identifying the true, safe minimum of what must be done to preserve the functions of physical assets. As a result, RCM has now been used by thousands of organizations spanning nearly every major field of organized human endeavor. It is becoming as fundamental to the practice of physical asset management as double-entry bookkeeping is to financial asset management.

The growing popularity of RCM has led to the development of numerous derivatives. Some of these derivatives are refinements and enhancements of Nowlan and Heap's original RCM process. However, less rigorous derivatives have also emerged, most of which are attempts to 'streamline' the maintenance strategy formulation process. This paper describes one of the most rigorous of the derivatives, known as RCM2. In order to place a RCM as a whole in context, the next section of this paper considers the recently-published SAE RCM Standard.[3]

## 2. THE SAE RCM STANDARD

As mentioned above, various derivatives of Nowlan and Heap's RCM process have emerged since their report was published in 1978. Many of these derivatives retain the key elements of the original process. However the widespread use of the term "RCM" led to the emergence of a number of processes that differ significantly from the original, but that their proponents also call "RCM." Many of these other processes either omit key steps of the process described by Nowlan and Heap, or change their sequence, or both. Consequently, despite claims to the contrary made by the proponents of these processes, the output differs markedly from what would be obtained by conducting a full, rigorous RCM analysis.

A growing awareness of these differences led to an increasing demand for a standard that set out the criteria any process must comply with in order to be called "RCM". Such a standard was published by the Society of Automotive Engineers (SAE) in 1999. An article[4] by Dana Netherton, Chairman of the SAE RCM Committee, described the evolution of RCM between 1978 and

1990, then went on to describe the evolution of the SAE Standard as quoted in the italicized paragraphs below:

***The Need for a Standard: the 1990s***

*Since the early 1990's, a great many more organizations have developed variations of the RCM process. Some, such as the US Naval Air Command with its 'Guidelines for the Naval Aviation Reliability Centered Maintenance Process (NAVAIR 00-25-403),[5] and the British Royal Navy with its RCM-oriented Naval Engineering Standard (NES45),[6] have remained true to the process originally expounded by Nowlan and Heap. However, as the RCM bandwagon has started rolling, a whole new collection of processes has emerged that are called "RCM" by their proponents, but that often bear little or no resemblance to the original meticulously researched, highly structured and thoroughly proven process developed by Nowlan and Heap. As a result, if an organization said that it wanted help in using or learning how to use RCM, it could not be sure what process would be offered.*

*Indeed, when the US Navy recently asked for Asset vendors to use RCM when building a new ship class, one US company offered a process closely related to the 1970 MSG-2 process. It defended its offering by noting that its process used a decision-logic diagram. Since RCM also uses a decision-logic diagram, the company argued, its process was an RCM process.*

*The US Navy had no answer to this argument, because in 1994 William Perry, the US Secretary of Defense, established a new policy about US military standards and specifications, which said that the US military would no longer require industrial vendors to use the military's 'standard' or 'specific' processes. Instead it would set performance requirements, and would allow vendors to use any processes that would provide Asset that would meet these requirements.*

*At a stroke, this voided the US military standards and specifications that defined "RCM." The US Air Force standard was cancelled in 1995. The US Navy has been unable to invoke its standards and specifications with Asset vendors (though it continues to use them for its internal work) – and it was unable to invoke them with the US company that wished to use MSG-2.*

*This development happened to coincide with the sudden interest in RCM in the industrial world. During the 1990s, magazines and conferences devoted to Asset maintenance have multiplied, and magazine articles and conference papers about RCM became more and more numerous. These have shown that very different processes are being given the same name, "RCM." So both the US military and commercial industry saw a need to define what an RCM process is.*

*In his 1994 memorandum, Perry said, "I encourage the Under Secretary of Defense (Acquisition and Technology) to form partnerships with industry associations to develop non-government standards for replacement of military standards where practicable." Indeed, the Technical Standards Board of the SAE has had a long and close relationship with the standards community in the US military, and has been working for several years to help develop commercial standards to replace military standards and specifications, when needed and when none already existed.*

*So in 1996 the SAE began working on an RCM-related standard, when it invited a group of representatives from the US Navy aviation and ship RCM communities to help it develop a standard for Scheduled Maintenance Programs. These US Navy representatives had already been meeting for about a year in an effort to develop a US Navy RCM process that might be common between the aviation and ship communities, so they had already done a considerable amount of work when they began to meet under SAE sponsorship. In late 1997, having gained members from commercial industry, the group realized that it was better to focus entirely on RCM. In 1998, the group found the best approach for its standard, and in 1999 it completed its draft of the standard, and the SAE approved it and published it.*

After a brief discussion about the practical difficulties associated with attempting to develop a universal standard of this nature, Netherton went on to say:

*"The standard now approved by the SAE does not present a standard process. Its title is, "Evaluation Criteria for Reliability-Centered Maintenance (RCM) Processes (SAE JA1011)." This standard presents criteria against which a process may be compared. If the process meets the criteria, it may confidently be called an "RCM process." If it does not, it should not. (This does not necessarily mean that processes that do not comply with the SAE RCM standard are not valid processes for maintenance strategy formulation. It simply means that the term "RCM" should not be applied to them.)"*

The italicized paragraph below quotes Section 5 of the Standard, which summarizes the key attributes of any RCM process as follows.

***Reliability-Centered Maintenance:*** *Any RCM process shall ensure that all of the following seven questions are answered satisfactorily and are answered in the sequence shown below:*

a. *What are the functions and associated desired standards of performance of the asset in its present operating context (functions)?*
b. *In what ways can it fail to fulfill its functions (functional failures)?*
c. *What causes each functional failure (failure modes)?*
d. *What happens when each failure occurs (failure effects)?*
e. *In what way does each failure matter (failure consequences)?*
f. *What should be done to predict or prevent each failure (proactive tasks and task intervals)?*
g. *What should be done if a suitable proactive task cannot be found (default actions)?*

*To answer each of the above questions "satisfactorily," the following information shall be gathered, and the following decisions shall be made. All information and decisions shall be documented in a way which makes the information and the decisions fully available to and acceptable to the owner or user of the asset.*

Subsequent sections of the Standard list the issues that any RCM process must address in order to answer each of the seven questions "satisfactorily." (In the rest of this paper, "true RCM" means any RCM process that complies with SAE Standard JA1011.)

In the mid-1980's, Aladon began applying RCM in industries other than aviation. From the technical point of view, the Aladon approach to RCM complies fully with SAE Standard

JA1011. The Aladon approach also pioneered the use of facilitated review groups to the application of true RCM, and the overall approach is known as RCM2. RCM2 is described in the book "Reliability-centered Maintenance" by John Moubray,[7] and it has now been applied on more than 1 200 industrial locations around the world. The rest of this paper describes RCM and RCM2 in more detail. Part 3 explores the meaning of the word 'maintenance', then goes on to define RCM and summarize key elements of the seven questions listed in SAE Standard JA1011. Part 4 describes how RCM2 should be applied, and Part 5 outlines what it achieves.

## 3. RCM

From the engineering viewpoint, there are two aspects to the management of any physical asset. It must be maintained and from time to time it may also need to be modified.

The major dictionaries define maintain as *cause to continue* (Oxford) or *keep in an existing state* (Webster). This suggests that maintenance means preserving something. On the other hand, they agree that to modify something means to change it in some way. The importance of this distinction is recognized in the RCM decision process. However, we focus on maintenance at this point.

When we set out to maintain something, what is it that we wish to cause to continue? What is the existing state that we wish to preserve?

The answer to these questions can be found in the fact that every physical asset is put into service because someone wants it do something. In other words, they expect it to fulfill a specific function or functions. So it follows that when we maintain an asset, the state we wish to preserve must be one in which it continues to do whatever its users want it to do.

***Maintenance: Ensuring that physical assets continue to do what their users want them to do***

What the users want depends on exactly where and how the asset is being used (the operating context). This leads to the following definition of Reliability-centered Maintenance:

***Reliability-centered Maintenance: a process used to determine what must be done to ensure that any physical asset continues to do whatever its users want it to do in its present operating context'.***

Before any version of true RCM can be applied to any asset or system, it is necessary to decide what system is to be analyzed, establish the system boundaries, clearly define its operating context, and prepare a detailed plan of action. These issues are discussed at greater length in Part 4 of this paper. This part of this paper briefly describes the RCM process itself.

The application of true RCM entails asking the seven questions set out in Part 2 above. The following sections of this paper summarize the key issues addressed when answering each of these questions.

### 3.1 Functions and Performance Standards

Before it is possible to determine what must be done to ensure that any physical asset continues to do whatever its users want it to do in its present operating context, we need to do two things:

- determine what its users want it to do
- ensure that it can do what its users want to start with.

This is why the first step in the RCM process entails defining the functions of each asset in its operating context, together with the desired standards of performance. What users expect assets to be able to do can be split into two categories:

- Primary functions, which summarizc why the asset was acquired in the first place. This category of functions covers issues such as speed, output, carrying or storage capacity, product quality and customer service.
- Secondary functions, which recognize that every asset is expected to do more than simply fulfill its primary functions. Users also have expectations in areas such as safety, control, containment, comfort, structural integrity, environmental compliance, economy, protection, efficiency of operation, and even the appearance of the asset.

## 3.2 Functional Failures

The objectives of maintenance are defined by the functions and associated performance expectations of the asset. But how does maintenance achieve these objectives?

The only occurrence that is likely to stop any asset performing to the standard required by its users is some kind of failure. This suggests that maintenance achieves its objectives by adopting a suitable approach to the management of failure. However, before we can apply a suitable blend of failure management tools, we need to identify what failures can occur. The RCM process does this at two levels:

- First, by identifying what circumstances amount to a failed state
- Then by asking what events can cause the asset to get into a failed state.

In the world of RCM, failed states are known as functional failures because they occur when an asset is unable to fulfill a function to a standard of performance which is acceptable to the user. In addition to the total inability to function, this definition encompasses partial failures, where the asset still functions but at an unacceptable level of performance (including situations where the asset cannot sustain acceptable levels of quality or accuracy).

## 3.3 Failure Modes

As mentioned in the previous paragraph, once each functional failure has been identified, the next step is to try to identify all the events that are reasonably likely to cause each failed state. These events are known as failure modes. 'Reasonably likely' failure modes include those that have occurred on the same or similar Asset operating in the same context, failures that are currently being prevented by existing maintenance regimes, and failures that have not happened yet but that are considered to be real possibilities in the context in question.

Most traditional lists of failure modes incorporate failures caused by deterioration or normal wear and tear. However, the list should include failures caused by human errors (on the part of operators and maintainers) and design flaws so that all reasonably likely causes of Asset failure can be identified and dealt with appropriately. It is also important to identify the cause of each failure in enough detail for it to be possible to identify an appropriate failure management policy.

### 3.4 Failure Effects

The fourth step in the RCM process entails listing failure effects, which describe what happens when each failure mode occurs. These descriptions should include all the information needed to support the evaluation of the consequences of the failure, such as:

- What evidence (if any) that the failure has occurred
- In what ways (if any) it poses a threat to safety or the environment
- In what ways (if any) it affects production or operations
- What physical damage (if any) is caused by the failure
- What must be done to repair the failure.

### 3.5 Failure Consequences

A detailed analysis of an average industrial undertaking is likely to yield between three and ten thousand possible failure modes. Each of these failures affects the organization in some way, but in each case, the effects are different. They may affect operations. They may also affect product quality, customer service, safety or the environment. They will all take time and cost money to repair.

It is these consequences which most strongly influence the extent to which we try to prevent each failure. In other words, if a failure has serious consequences, we are likely to go to great lengths to try to avoid it. On the other hand, if it has little or no effect, then we may decide to do no routine maintenance beyond basic cleaning and lubrication.

A great strength of RCM is that it recognizes that the consequences of failures are far more important than their technical characteristics. In fact, it recognizes that the only reason for doing any kind of proactive maintenance is not to avoid failures per se, but to avoid or at least to reduce the consequences of failure. The RCM process classifies these consequences into four groups, as follows:

- *Hidden failure consequences:* Hidden failures have no direct impact, but they expose the organization to multiple failures with serious, often catastrophic, consequences.
- *Safety and environmental consequences:* A failure has safety consequences if it could injure or kill someone. It has environmental consequences if it could breach a corporate, regional, national or international environmental standard.
- *Operational consequences:* A failure has operational consequences if it affects operations (output, product quality, customer service or operating costs in addition to the direct cost of repair)
- *Non-operational consequences:* Evident failures that fall into this category affect neither safety nor production, so they involve only the direct cost of repair.

The RCM process uses these categories as the basis of a strategic framework for maintenance decision-making. By forcing a structured review of the consequences of each failure mode in terms of the above categories, it integrates the operational, environmental and safety objectives

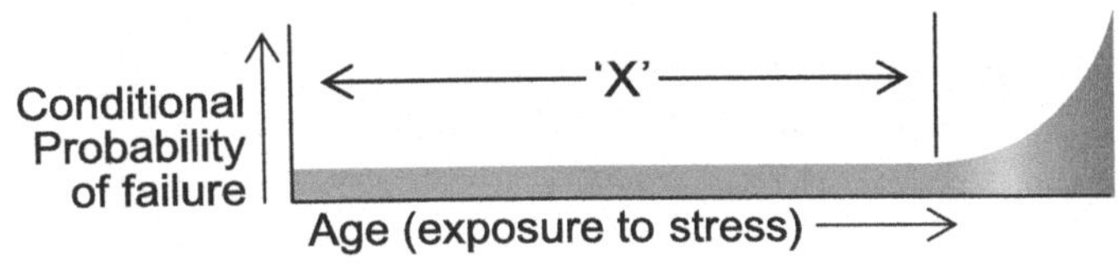


**Figure D.1 The traditional view of failure**

of maintenance. This helps to bring safety and the environment into the mainstream of maintenance management.

The consequence evaluation process also shifts emphasis away from the idea that all failures are bad and must be prevented. In so doing, it focuses attention on the maintenance activities that have most effect on the performance of the organization, and diverts energy away from those that have little or no effect. It also encourages us to think more broadly about different ways of managing failure, rather than to concentrate only on failure prevention. Failure management techniques are divided into two categories:

- Proactive tasks: these are tasks undertaken before a failure occurs, in order to prevent the item from getting into a failed state. They embrace what is traditionally known as "predictive" and "preventive" maintenance, although RCM uses the terms scheduled restoration, scheduled discard and on-condition maintenance.
- Default actions: these deal with the failed state, and are chosen when it is not possible to identify an effective pro-active task. Default actions include failure-finding, redesign and run-to-failure.

## 3.6 Proactive Tasks

Many people still believe that the best way to optimize plant availability is to do some kind of proactive maintenance on a routine basis. Conventional wisdom suggested that this should consist of fixed interval overhauls or component replacements. Figure D.1 illustrates the fixed interval view of failure.

Figure D.1 is based on the assumption that most items operate reliably for a period 'X,' and then wear out. Classical thinking suggests that extensive records about failure will enable us to determine this life and so make plans to take preventive action shortly before the item is due to fail in future.

This model is true for certain types of simple Asset, and for some complex items with dominant failure modes. In particular, wear-out characteristics are often found where Asset comes into direct contact with the product. Age-related failures are also often associated with fatigue, corrosion, abrasion and evaporation.

However, Asset in general is far more complex than it was thirty years ago. This has led to startling changes in the patterns of failure, as shown in Figure D.2. The graphs show conditional probability of failure against operating age for a variety of electrical and mechanical items.

- Pattern A is the well-known bathtub curve. It begins with high incidence of failure (known as infant mortality) followed by constant or gradually increasing conditional probability of failure, then by a wear-out zone

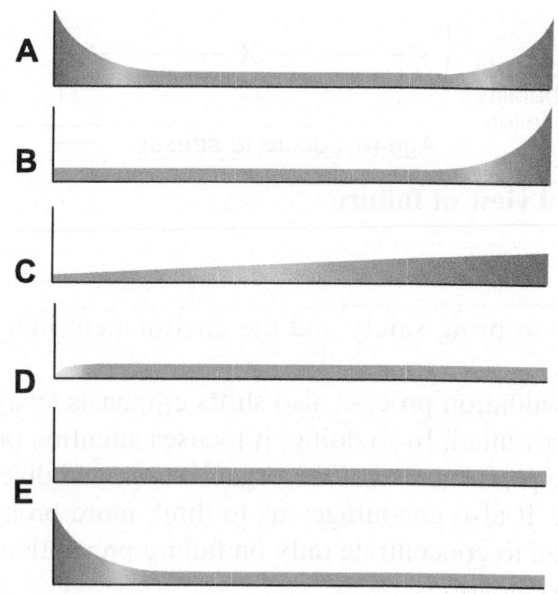


**Figure D.2 The reality of failure**

- Pattern B shows constant or slowly increasing conditional probability of failure, ending in a wear-out zone (the same as Figure D.1).
- Pattern C shows slowly increasing conditional probability of failure, no identifiable wear-out age.
- Pattern D shows low conditional probability of failure when the item is new or just out of the shop, then a rapid increase to a constant level.
- Pattern E shows a constant conditional probability of failure at all ages (random failure).
- Pattern F starts with high infant mortality, dropping to a constant or slowly decreasing conditional probability of failure.

Studies done on civil aircraft showed that 4% of the items conformed to pattern A, 2% to B, 5% to C, 7% to D, 14% to E and no fewer than 68% to pattern F. (The number of times that these patterns occur in aircraft is not necessarily the same as in industry. But there is no doubt that as assets become more complex, we see more and more of patterns E and F.)

These findings contradict the belief that there is always a connection between reliability and operating age. This belief led to the idea that the more often an item is overhauled, the less likely it is to fail. Nowadays, this is seldom true. Unless there is a dominant age-related failure mode, age limits do little or nothing to improve the reliability of complex items. In fact scheduled overhauls often increase overall failure rates by introducing infant mortality into otherwise stable systems. An awareness of these facts has led some organizations to abandon the idea of proactive maintenance completely. In fact, this can be the right thing to do for failures with minor consequences. But when the failure consequences are significant, something must be done to prevent or predict the failures, or at least to reduce the consequences.

This brings us back to the question of proactive tasks. As mentioned earlier, RCM divides proactive tasks into three categories, as follows:

- scheduled restoration tasks
- scheduled discard tasks
- scheduled on-condition tasks.

### *Scheduled Restoration and Scheduled Discard Tasks*

Scheduled restoration entails remanufacturing a component or overhauling an assembly at or before a specified age limit, regardless of its condition at the time. Similarly, scheduled discard entails discarding an item at or before a specified life limit, regardless of its condition at the time.

Collectively, these two types of tasks are now generally known as preventive maintenance. They used to be by far the most widely used form of proactive maintenance. However for the reasons discussed above, they are much less widely used than they were twenty years ago.

### *On-Condition Tasks*

The continuing need to prevent certain types of failure, and the growing inability of classical techniques to do so, are behind the growth of new types of failure management. The majority of these techniques rely on the fact that most failures give some warning of the fact that they are about to occur. These warnings are known as potential failures, and are defined as identifiable physical conditions which indicate that a functional failure is about to occur or is in the process of occurring.

The new techniques are used to detect potential failures so that action can be taken to reduce or eliminate the consequences that could occur if they were to degenerate into functional failures. They are called on-condition tasks, and include all forms of condition-based maintenance, predictive maintenance and condition monitoring.)

Used appropriately, on-condition tasks are a very good way of managing failures, but they can also be an expensive waste of time. RCM enables decisions in this area to be made with particular confidence.

## 3.7 Default Actions

RCM recognizes three major categories of default action:

- *Failure-finding:* Failure-finding entails checking hidden functions to find out if they have failed (as opposed to the on-condition tasks described above, which entail checking if something is failing). The rapid growth in the use of built-in protective devices means that this category of tasks is likely to become as big a maintenance management issue in the next ten years as condition monitoring has been in the last decade. RCM provides powerful, risk-based rules for establishing whether, how often and by whom these tasks should be done.
- *Redesign:* redesign entails making any one-time change to the built-in capability of a system. This includes modifications to hardware and changes to procedures. (Note that the RCM process considers the maintenance requirements of each asset before asking whether it is necessary to change the design. This is because the maintenance person who is on duty today has to maintain the asset as it exists today, not what should be there or what might be there at some stage in the future. However, if it transpires that

an asset simply cannot deliver the desired performance, RCM helps to focus redesign efforts on the real problems).

- *No scheduled maintenance:* as the name suggests, this default entails making no effort to anticipate or prevent failure modes to which it is applied, and so those failures are simply allowed to occur and then repaired. This default is also called run-to-failure.

### 3.8 The RCM Task Selection Process

A great strength of RCM is the way it provides precise and easily understood criteria for deciding which (if any) of the proactive tasks is technically feasible in any context, and if so for deciding how often and by whom they should be done. Whether or not a proactive task is technically feasible is governed by the technical characteristics of the task and of the failure that it is meant to prevent. Whether it is worth doing is governed by how well it deals with the consequences of the failure. If a proactive task cannot be found that is both technically feasible and worth doing, then suitable default action must be taken. The essence of the task selection process is as follows:

- For *hidden failures*, a proactive task is worth doing if it reduces the risk of the multiple failure associated with that function to a tolerably low level. If such a task cannot be found then a scheduled failure-finding task must be pre-scribed. If a suitable failure-finding task cannot be found, then the secondary default decision is that the item may have to be redesigned (depending on the consequences of the multiple failure).
- For failures with *safety or environmental consequences*, a proactive task is only worth doing if it reduces the risk of that failure on its own to a very low level indeed, if it does not eliminate it altogether. If a task cannot be found that reduces the risk of the failure to a tolerable level, the item must be redesigned or the process must be changed.
- If the failure has *operational consequences*, a proactive task is only worth doing if the total cost of doing it over a period of time is less than the cost of the operational consequences and the cost of repair over the same period. In other words, the task must be justified on economic grounds. If it is not justified, the initial default decision is no scheduled maintenance. (If this occurs and the operational consequences are still unacceptable then the secondary default decision is again redesign).
- If a failure has *non-operational consequences* a proactive task is only worth doing if the cost of the task over a period of time is less than the cost of repair over the same period. So these tasks must also be justified on economic grounds. If it is not justified, the initial default decision is again no scheduled maintenance, and if the repair costs are too high, the secondary default decision is once again redesign.

This approach means that proactive tasks are only specified for failures that really need them, which in turn leads to substantial reductions in routine workloads. Less routine work also means that the remaining tasks are more likely to be done properly. This together with the elimination of counterproductive tasks leads to more effective maintenance.

Compare this with the traditional approach to the development of maintenance policies. Traditionally, the maintenance requirements of each asset are assessed in terms of its real or assumed technical characteristics, with-out considering the consequences of failure. The resulting schedules are used for all similar assets, again without considering that different consequences apply in different operating contexts. This results in large numbers of schedules that are wasted, not because they are 'wrong' in the technical sense, but because they achieve nothing.

Correctly applied, RCM leads to remarkable improvements in maintenance effectiveness, and often does so surprisingly quickly. However, as with any fundamental change management project, RCM is much more likely to succeed if proper attention is paid to thorough planning, how and by whom the analysis is performed, auditing and implementation. These issues are discussed in the following paragraphs.

## 4. APPLYING RCM2

### Prioritizing Assets and Establishing Objectives

Part 5 of this paper explains that RCM can improve organizational performance in a host of different ways, tangible and intangible. Tangible benefits include greater safety, improved environmental integrity, improved availability and reliability, better product quality and customer service and reduced operating and maintenance costs. Intangible benefits include better understanding about how the Asset works on the part of operators and maintainers, improved teamworking and higher morale.

RCM should be applied first to systems where it is likely to yield the highest returns relative to the effort required in any or all of the above areas. If these systems are not self-evident, it may be necessary to prioritize RCM projects on a more formal basis. When this has been done, it is then essential to plan each project in detail.

### Planning

The successful application of RCM depends first and perhaps foremost on meticulous planning and preparation. The key elements of the planning process are as follows:

- Define the scope and boundaries of each project
- Define and wherever possible quantify the objectives of each project (now state and desired end state)
- Estimate the amount of time (number of meetings) needed to review the Asset in each area
- Identify project manager and facilitator(s)
- Identify participants (by title and by name)
- Plan training for participants and facilitators
- Plan date, time and location of each meeting
- Plan management audits of RCM recommendations
- Plan to implement the recommendations (maintenance tasks, design changes, changes to operating procedures)

### Review Groups

We have seen how the RCM process embodies seven basic questions. In practice, maintenance people simply cannot answer all these questions on their own. This is because many (if not

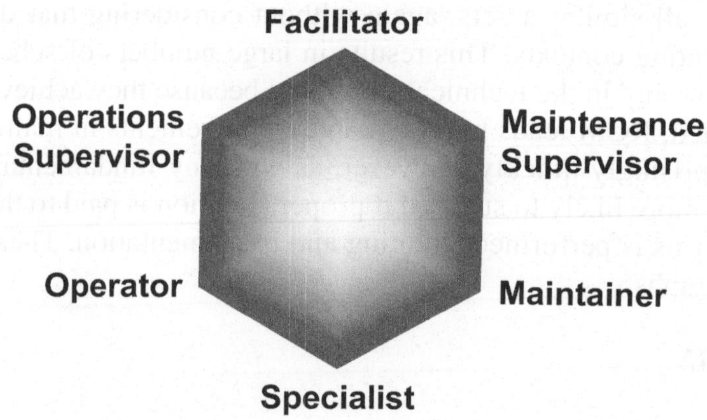


**Figure D.3 A typical RCM2 review group**

most) of the answers can only be supplied by production or operations people. This applies especially to questions concerning functions, desired performance, failure effects and failure consequences.

For this reason, a review of the maintenance requirements of any asset should be done by small teams that include at least one person from the maintenance function and one from the operations function. The seniority of the group members is less important than the fact that they should have a thorough knowledge of the asset under review. Each group member should also have been trained in RCM. The make-up of a typical RCM2 review group is shown in Figure D.3.

The use of these groups not only enables management to gain access to the knowledge and expertise of each member of the group on a systematic basis, but the members themselves learn a great deal about how the asset works.

**Facilitators**

RCM review groups work under the guidance of highly trained specialists in RCM, known as facilitators. The facilitators are the most important people in the RCM review process. Their role is to ensure that:

- The RCM analysis is carried out at the right level, that system boundaries are clearly defined, that no important items are overlooked and that the results of the analysis are properly recorded
- RCM is correctly understood and applied by the group
- The group reaches consensus in a brisk and orderly fashion, while retaining their enthusiasm and commitment
- The analysis progresses as planned and finishes on time.

Facilitators also work with RCM project managers or sponsors to ensure that each analysis is properly planned and receives appropriate managerial and logistic support.

### *The Outcomes of an RCM Analysis*

If it is applied in the manner suggested above, an RCM analysis results in three tangible outcomes, as follows:

- schedules to be done by the maintenance department
- revised operating procedures for the operators of the asset
- A list of areas whcre one-time changes must be made to the design of the asset or the way it is operated to deal with situations where the asset cannot deliver the desired performance in its current configuration.

A less tangible but very valuable outcome is that participants in the process tend to start functioning much better as multi-disciplinary teams after their analyses have been completed.

### Auditing

After the review has been completed for each asset, senior managers with overall responsibility for the Asset must satisfy themselves that the review is sensible and defensible. This entails deciding whether they agree with the definition of functions and performance standards, the identification of failure modes and the description of failure effects, the assessment of failure consequences and the selection of tasks.

### Implementation

Once the RCM review has been audited and approved, the final step is to implement the tasks, procedures and one-time changes. The revised tasks and procedures must be documented in a way that ensures that they will be easily under-stood and performed safely by the people who do the work.

The maintenance tasks are then fed into suitable planning and scheduling systems, while revised operating procedures are usually incorporated into standard operating procedure manuals. Proposals for modifications are dealt with by the engineering or project management function in most organizations.

## 5. WHAT RCM2 ACHIEVES

Desirable as they are, the outcomes listed above should only be seen as a means to an end. Specifically, they should enable the maintenance function to fulfill all the expectations listed in Figure D.1 at the beginning of this paper. How they do so is summarized in the following paragraphs.

- *Greater safety and environmental integrity:* RCM considers the safety and environmental implications of every failure mode before considering its effect on operations. This means that steps are taken to minimize all identifiable Asset-related safety and environmental hazards, if not eliminate them altogether. By integrating safety into the mainstream of maintenance decision-making, RCM also improves attitudes to safety.

- *Improved operating performance (output, product quality and customer service):* RCM recognizes that all types of maintenance have some value, and provides rules for deciding which is most suitable in every situation. By doing so, it helps ensure that only the most effective forms of maintenance are chosen for each asset, and that suitable action is taken in cases where maintenance cannot help. This much more tightly focused maintenance effort leads to quantum jumps in the performance of existing assets where these are sought.

  RCM was developed to help airlines draw up maintenance programs for new types of aircraft before they enter service. As a result, it is an ideal way to develop such programs for new assets, especially complex Asset for which no historical information is available. This saves much of the trial and error that is so often part of the development of new maintenance programs – trial that is time-consuming and frustrating, and error that can be very costly.
- *Greater maintenance cost-effectiveness:* RCM continually focuses attention on the maintenance activities that have most effect on the performance of the plant. This helps to ensure that everything spent on maintenance is spent where it will do the most good.

  In addition, if RCM is correctly applied to existing maintenance systems, it reduces the amount of routine work (in other words, maintenance tasks to be undertaken on a cyclic basis) issued in each period, usually by 40% to 70%. On the other hand, if RCM is used to develop a new maintenance program, the resulting scheduled workload is much lower than if the program is developed by traditional methods.
- *A comprehensive database:* An RCM review ends with a comprehensive and fully documented record of the maintenance requirements of all the significant assets used by the organization. This makes it possible to adapt to changing circumstances (such as changing shift patterns or new technology) without having to reconsider all maintenance policies from scratch. It also enables Asset users to demonstrate that their maintenance programs are built on rational foundations (the audit trail required by more and more regulators). Finally, the information stored on RCM worksheets reduces the effects of staff turnover with its attendant loss of experience and expertise.

  An RCM review of the maintenance requirements of each asset also provides a much clearer view of the skills required to maintain each asset, and for deciding what spares should be held in stock.
- *Greater motivation of individuals,* especially people who are involved in the review process. This is accompanied by much wider 'ownership' of maintenance problems and their solutions. It also means that solutions are more likely to endure.
- *Better teamwork:* RCM provides a common, easily understood technical language for everyone who has anything to do with maintenance. This gives maintenance and operations people a better understanding of what maintenance can (and cannot) achieve and what must be done to achieve it.

All of these issues are part of the mainstream of maintenance management, and many are already the target of improvement programs. A major feature of RCM is that it provides an effective step-by-step framework for tackling all of them at once, and for involving everyone who has anything to do with the Asset in the process.

RCM yields results very quickly. In fact, if they are correctly focused and correctly applied, RCM analyses can pay for themselves in a matter of months and sometimes even a matter of weeks. The process transforms both the perceived maintenance requirements of the physical assets used by the organization and the way in which the maintenance function as a whole is perceived. The result is more cost-effective, more harmonious and much more successful maintenance.

## REFERENCES


1. Nowlan FS and Heap H: "Reliability-Centered Maintenance." Springfield, Virginia. National Technical Information Service, United States Department of Commerce.
2. Maintenance Steering Group - 3 Task Force: "Maintenance Program Development Document MSG-3." Washington DC: Air Transport Association (ATA) of America, 1993.
3. International Society of Automotive Engineers: "JA1011 - Evaluation Criteria for Reliability-Centered Maintenance (RCM) Processes." Warrendale, Pennsylvania, USA: SAE Publications.
4. Netherton D: "SAE's New Standard for RCM." *Maintenance* (UK) 15 (1) 3-7, 2000.
5. US Naval Air Systems Command: "NAVAIR 00-25-403: Guidelines for the Naval Aviation Reliability Centered Maintenance Process." Philadelphia, Pennsylvania. US Department of Defense Publications.
6. RCM Implementation Team, Royal Navy: "NES 45 Naval Engineering Standard 45, Requirements for the Application of Reliability-Centered Maintenance Techniques to HM Ships, Royal Fleet Auxiliaries and other Naval Auxiliary Vessels." Foxhill, Bath, United Kingdom. UK Ministry of Defence Publications.
7. Moubray JM: "Reliability-Centered Maintenance": New York, New York USA: Industrial Press.

# APPENDIX E
# RCM PROCESS FLOW AND DECISION DIAGRAM

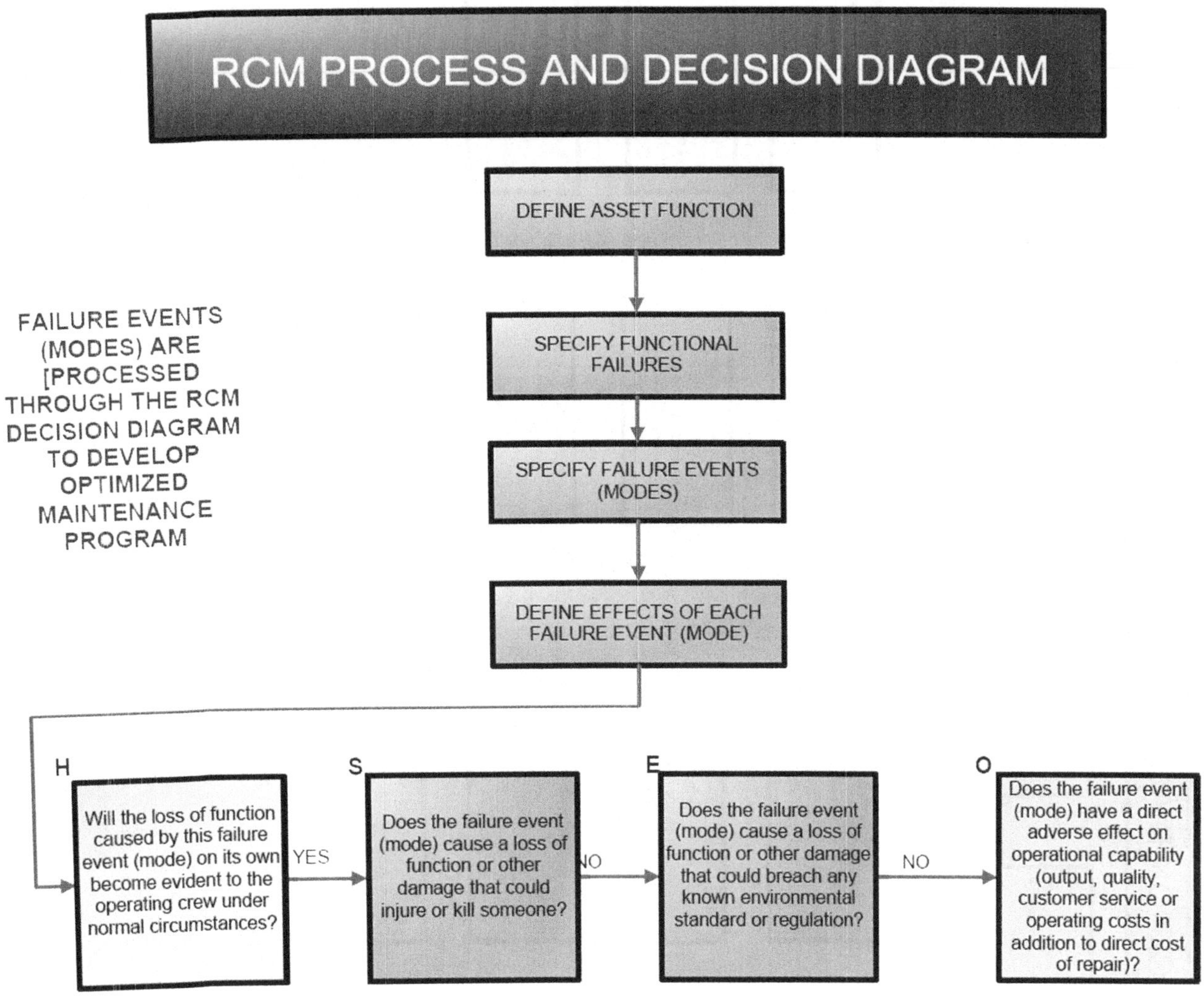


**Figure E.1 RCM process overview**

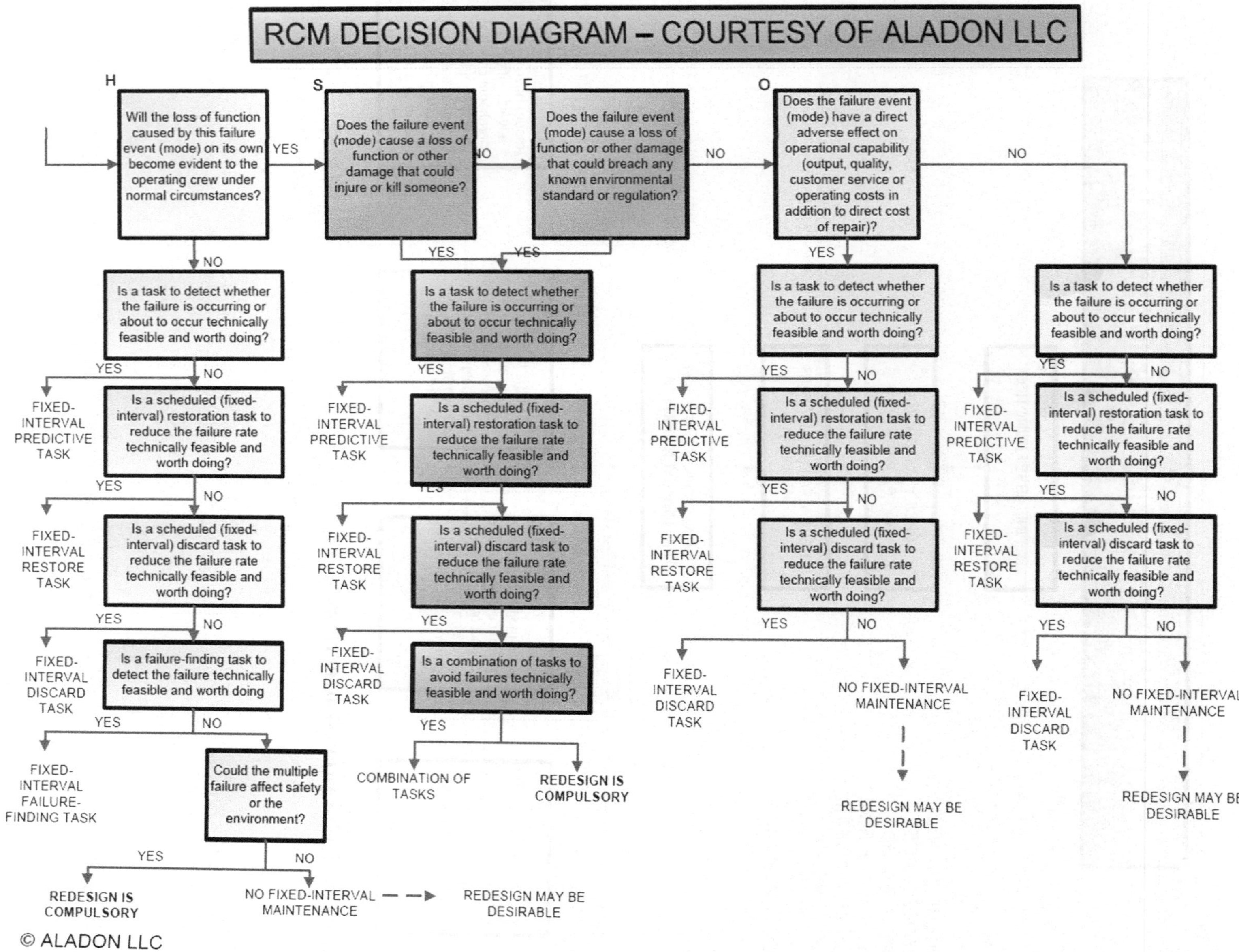


*Source:* Courtesy of Aladon LLC

**Figure E.2 RCM2 decision diagram**

# GLOSSARY

**Abandoned** Permanently removed from service

**Abnormal conditions** Conditions that exceed or are expected to exceed operating design, limits or events which are considered abnormal conditions, even if operating design limits have not been exceeded.

**Check valves** Operates one-way and prevents the reverse flow of fluid. Check valves are designed to be held open by flowing liquid and to drop closed automatically when flow stops or is reversed.

**CMMS** Computerized Maintenance Management System. It is a maintenance tracking software system and it is part of managing and controlling assets, plant and equipment maintenance.

**Code** A code is an Industry standard or guide for design, construction, operation and maintenance of a pipeline and/or pipeline facilities. It may be incorporated wholly as in part into a regulation.

**Collar** Screwed couplings on a shaft

**Conditional probability of failure** The probability that a failure will occur in a specific period provided that the item concerned has survived to the beginning of that period.

**Desired performance** The level of performance acceptable to the owner or user of the physical asset or system.

**Discharge pressure** Pressure of the fluid (liquid) as it exits a pump.

**Down-time** Loss of primary function for asset system

**Enterprise Asset Management (EAM)** Enterprise Asset Management [EAM] refers to the systems, data, people and processes required to manage physical assets through their full life-cycle across all enterprise business functions in order to maximize enterprise ROI over the asset life-cycle.

**Environmental consequences** A failure mode or a multiple failure has environmental consequences if it could breach any corporate, municipal, regional, national or international environmental standard or regulation which applies to the physical asset or system under consideration.

**Eutrophic** Having waters rich in mineral and organic nutrients that promote a proliferation of plant life, especially algae, which reduces the dissolved oxygen content and often causes the extinction of other organisms. Used of a lake or pond.

**Eutrophication** Aging of a lake by biological enrichment of its water. In a young lake the water is cold and clear, supporting little life. With time, streams draining into the lake introduce nutrients such as nitrogen and phosphorus, which encourage the growth of aquatic organisms. As the lake's fertility increases, plant and animal life burgeons, and organic remains begin to be deposited on the lake bottom. Over the centuries, as silt and organic debris pile up, the lake grows shallower and warmer, with warm-water organisms supplanting those that thrive in a cold environment. Marsh plants take root in the shallows and begin to fill in the original lake basin. Eventually the lake gives way to *bog*, finally disappearing into land. Depending on climate, size of the lake, and other factors, the natural aging of a lake may span thousands of years. However, pollutants from man's activities can radically accelerate the aging process. During the past century, lakes in many parts of the earth have been severely eutrophied by sewage and agricultural and industrial wastes (see *water pollution*). The prime contaminants are nitrates and phosphates, which act as plant nutrients. They over stimulate the growth of algae, causing unsightly scum and unpleasant odors, and robbing the water of dissolved oxygen vital to other aquatic life. At the same time, other pollutants flowing into a lake may poison whole populations of fish, whose decomposing remains further deplete the water's dissolved oxygen content. In such fashion, a lake can literally choke to death.

**Evident failure** Failure modes that will on their own become evident to operators under normal operating conditions.

**Evident function** A function whose failure on its own will become evident to the operating crew under normal operating conditions.

**Facilitator** Facilitators are trained in the RCM methodology and mentored by external consultants to be competent in 45 RCM skill sets.

**Facility** Refer to Water Treatment facility or Treatment System.

**Failure consequences** The way (or ways) in which a failure or a multiple failure matter

**Failure effect** The physical effects of a failure mode when it happens and nothing are done to prevent it from happening.

**Failure finding task** A scheduled task to seek to determine whether a specific hidden failure has occurred (check whether an item has failed)

**Failure management policy** A generic term that encompasses on-condition tasks, scheduled restoration, scheduled discard, failure finding, run-to-failure and redesign.

**Failure mode** A single event that causes a functional failure

**Function** What the owner or user of a physical asset or system want it to do

**Functional failure** A state in which a physical asset or system is unable to perform a specific function to a level of performance that is acceptable to its owner or user.

**Hazardous condition** A situation that could result in harm or injury to persons or could severely damage property or the environment

**Heat affected zone** The portion of the base metal that has not been melted but whose mechanical properties or microstructures has been altered by the heat of welding or cutting

**Hidden failure** A failure that will not become evident to the operating crew when it occurs on its own under normal circumstances.

**Hidden function** A function whose failure will not become evident to the operating crew under normal circumstances if it occurs on its own

**High-side pumps** Centrifugal Pumps delivering water to Belleview reservoir at elevation 5743'

**Initial capability** The level of performance of which a physical asset or system is capable at the moment it enters service.

**KPI** Key metric for measuring performance

**Local operation** Control and operation of equipment and systems from the equipment location (field facilities).

**Low-side pumps** Centrifugal Pumps delivering water to Kendrick reservoir at elevation 5627'

**Maximum allowable operating pressure (MAOP)** A rating indicating the maximum pressure at which a pipeline or segment of a pipeline may be operated under regulations in normal conditions. Also called pressure rating.

**Monitoring** Acquisition, storage and evaluation of quantitative data using specific instrumentation. Monitoring information is gathered and used to recognize trends, detect unknown and unexpected problems, and plan and prioritize repair work.

**MTTR** Mean-time-to-repair – average time to perform a defined maintenance task or elapsed time to restore required function of an asset.

**Multiple failure** An event that occurs if a protected function fails while its protective device or system is in a failed state.

**NDE** Nondestructive examination – any testing method which does not involve damaging or destroying the test sample; includes use of X-ray, ultrasonic, radiography, magnetic flux, etc.

**Nominal wall thickness** The wall thickness listed in the pipe specifications

**Non conforming conditions** An item that does not conform to established acceptance criteria or conformance to established acceptance criteria is indeterminate.

**Non-operational consequences** A failure mode has non-operational consequences if it is not hidden and does not have safety, environmental or operational consequences, but only requires repair.

**Normal operations** Operating the Treatment facility within design limits

**Notable conditions** New or significantly changed condition, which require repairs beyond the scope of routine maintenance or local resources and/or may require an engineering review.

**On-condition task** A scheduled task used to determine whether a potential failure has occurred or is in the process of happening. (Check whether an item is failing or about to fail.)

**Operating context** The circumstances in which a physical asset or system is expected to operate.

**Operational consequences** A failure mode or multiple failure has operational consequences if it could adversely affect the operational capability of a physical asset or system (output, product quality, customer service, military capability, or operating costs in addition to the cost of repair).

**Operator** Any individual, firm, joint venture, partnership, corporation, association, State, municipality, cooperative association, or joint stock association and includes any trustee, receiver, assignee, or personal representative thereof who engages in the treatment of waste water.

**P-F interval** The intervals between the point at which a potential failures becomes detectable and the point at which it degrades into a functional failure (also known as 'failure development period' or 'lead time to failure'

**Person** Any individual, firm, joint venture, partnership, corporation, association, state, municipality, cooperative association, or joint stock association and includes any trustee, receiver, assignee, or personal representative thereof

**Pilot project** Initial RCM project performed on an asset or assets

**Planning** The process of defining the scope of work, required logistics and major operations and sequencing required for a maintenance task.

**Portal** Single point of access to multiple systems and sources of data

**Potential failure** An identifiable condition that indicates that a functional failure is either about to occur or is in the process of occurring.

**Pressure relief valve (PRV)** A valve designed to open automatically to relieve pressure and keep it below a designated level.

**Primary function** The function(s) that constitutes the main reason(s) why a physical asst or system is acquired by its owner or user.

**Procedure** A document that specifies or describes how an activity is to be performed

**Protective device or protective system** A device or system that is intended to avoid, eliminate or minimize the consequences of failure of some other system.

**Qualified welder** A welder who has demonstrated his ability to produce welds meeting prescribed standards.

**Qualified welding procedure** A tested and proven detailed method by which sound welds having suitable mechanical properties can be produced.

**Quench rapid cooling** Quenching of the heat-affected zone of a weld may cause increased hardness and brittleness and reduced ductility and toughness.

**RCM** Process for determining maintenance requirements of an asset in its current operating context

**Redesign** Any action taken to change the physical configuration of an asset or system (modification), to change the operating context of the asset or system, the change the method used by an operator or maintainer to perform a task, or to change the capability of an operator or maintainer (training).

**Regulation** A published code, standard or rule prescribing minimum requirements and having full effect of law.

**Remote operation** Control of the Waste Water Treatment system operations from the Operations Control Room.

**Routine** Performed at predetermined and fixed intervals.

**Run-to-failure** A failure management policy that permits a specific failure mode to occur without any attempt to anticipate or prevent it

**Safety consequences** A failure mode or a multiple failure has safety consequences if it could injure or kill a human being.

**Scheduled** Performed at predetermined and fixed intervals.

**Scheduled discard** A task that entails discarding (replacing) an item or component at or before a specific age limit regardless of its condition at the time.

**Scheduled restoration** A task that restores (repairs) the initial capability of an item or component at or before a specific age limit regardless of its condition at the time.

**Scheduling** The process of assigning planned maintenance tasks to specific time buckets or periods and crews.

**Secondary function** Functions which a physical asset or system has to fulfill apart from its primary function(s), such as those required to fulfill regulatory requirements, and those which concern issues such as protection, control, containment, comfort, appearance, structural integrity and energy efficiency.

**Skill set** To be competent in a specific area required to be a RCM facilitator

# REFERENCES


Bloom, Neil B. 2005. *Reliability Centered Maintenance—Implementation Made Simple*. New York: McGraw-Hill

Hefner, R., and Smith, A.M. 2002. *The Application of RCM to Optimizing a Coal Pulverizer Preventative Maintenance Program*. Presentation to SMRP Annual Conference, October 2002, Nashville, Tennessee.

International Society of Automotive Engineers. 1999. *JA1011—Evaluation Criteria for Reliability-Centered Maintenance (RCM) Processes*. Warrendale, Pa.: SAE Publications.

Maintenance Steering Group-3 Task Force. 1993. *Maintenance Program Development Document MSG-3*. Washington, D.C.: Air Transport Association of America.

Moubray, J. 1997. *Reliability Centered Maintenance II*. New York: Industrial Press.

Netherton, D. 2000. SAE's New Standard for RCM. *Maintenance*, 15(1):3–7.

Nowlan, F.S., and Heap, H. 1978. *Reliability-Centered Maintenance*. Springfield, Va.: U.S. Department of Commerce.

RCM Implementation Team, Royal Navy. 2000. *NES 45 Naval Engineering Standard 45, Requirements for the Application of Reliability-Centered Maintenance Techniques to HM Ships, Royal Fleet Auxiliaries and Other Naval Auxiliary Vessels*. Foxhill, Bath, UK: U.K. Ministry of Defence Publications.

Smith, A.M. 1993. *Reliability-Centered Maintenance*. New York: McGraw-Hill.

Smith, A.M., and Hinchcliffe, G.R. 2004. *RCM—Gateway to World Class Maintenance*. Burlington, Mass.: Elsevier Butterworth-Heinemann.

U.S. Naval Air Systems Command. 2003. *NAVAIR 00-25-403: Guidelines for the Naval Aviation Reliability Centered Maintenance Process* [Online]. Available from http://logistics.navair.navy.mil/rcm/library/403manual.pdf.

Veolia Water Indianapolis LLC. 2004. *Annual Report, City of Indianapolis, Indiana*. Indianapolis, Ind. [Online]. Available from http://www.indianapoliswater.com/Assets/PDF/VWI_Annual_Report.pdf.

# ABBREVIATIONS

| | |
|---|---|
| ANSI | American National Standards Institute |
| ASME | American Society of Mechanical Engineers |
| AwwaRF | Awwa Research Foundation |
| CMMS | computerized maintenance management system |
| DWB | Denver Water Board |
| EAM | enterprise asset management |
| GASB | governmental accounting standards board |
| HAZOP | hazards and operability analysis |
| ID | inside diameter |
| MGD | millions of gallons per day |
| NDS | New Dimension Solutions, Inc |
| NPS | nominal pipe size |
| OD | outside diameter |
| OEM | original equipment manufacturer |
| OSHA | Occupational Safety and Health Administration |
| PdM | predictive maintenance |
| PM | preventive maintenance |
| PSH | pressure switch high |
| PSR | pipeline surveillance report |
| PSV | pressure safety valve |
| RCM | reliability-centered maintenance |
| RTD | resistance temperature device |
| SCADA | supervisory control and data acquisition |
| SIP | standard inspection procedure |
| SME | subject matter expert |
| SMP | safe maintenance procedure |
| SOP | safe operating procedure |
| SPOC | single point of contact |
| SPP | safe project procedure |
| TPM | total productive maintenance |

IWA Publishing's authorised EU representative for General Product Safety Regulations is Diane D'Arras, 15 rue Duret, 75116 Paris, France, e-mail: safety@iwap.co.uk.

Printed and bound by CPI Group (UK) Ltd, Croydon, CR0 4YY
17/09/2026
02218811-0002